Statistics and Pattern Recognition Applied to the Spatio-Temporal Properties of Seismicity

Statistics and Pattern Recognition Applied to the Spatio-Temporal Properties of Seismicity

Editors

Stefania Gentili
Rita Di Giovambattista
Robert Shcherbakov
Filippos Vallianatos

MDPI • Basel • Beijing • Wuhan • Barcelona • Belgrade • Manchester • Tokyo • Cluj • Tianjin

Editors
Stefania Gentili
National Institute of
Oceanography and Applied
Geophysics - OGS
Italy

Rita Di Giovambattista
Istituto NAzionale di
Geofisica e Vulcanologia
(INGV)
Italy

Robert Shcherbakov
Western University
Canada

Filippos Vallianatos
National and Kapodistrian
University of Athens
Greece

Editorial Office
MDPI
St. Alban-Anlage 66
4052 Basel, Switzerland

This is a reprint of articles from the Special Issue published online in the open access journal *Applied Sciences* (ISSN 2076-3417) (available at: https://www.mdpi.com/journal/applsci/special_issues/Pattern_Recognition_Seismicity).

For citation purposes, cite each article independently as indicated on the article page online and as indicated below:

LastName, A.A.; LastName, B.B.; LastName, C.C. Article Title. *Journal Name* **Year**, *Volume Number*, Page Range.

ISBN 978-3-0365-4263-8 (Hbk)
ISBN 978-3-0365-4264-5 (PDF)

Cover image courtesy of Robert Shcherbakov.

Contents

Preface to "Statistics and Pattern Recognition Applied to the Spatio-Temporal Properties of Seismicity"

In recent years, there have been significant advances in the understanding of seismicity scaling laws, the study of spatiotemporal correlations, and earthquake clustering, with direct implications for time-dependent seismic hazard assessment. New models based on seismicity patterns, considering their physical and statistical significance, have shed light on the preparation process before large earthquakes and the evolution of clustered seismicity in time and space. On the other hand, the increasing amount of seismic data available at both local and global scales, together with accurate assessments of the reliability of the catalogs, offers new opportunities for model verification.

This Special Issue brings together eight peer-reviewed articles. The articles represent a collection of innovative applications of earthquake forecasting, including the earthquake preparation process, seismic hazard assessment, statistical analysis of seismicity, synthetic catalogs, and cluster identification.

It is therefore invaluable to seismologists, statistical seismologists, research students, government agencies, and academics.

We are especially grateful to all the authors as without them this Special Issue would not have become a reality. As guest editors, we would like to thank the reviewers for their careful evaluation and valuable contributions. Special thanks go to Assistant Editors Carlos Sanchez and Jill Fang for their dedication to this project and their invaluable collaboration in setting up, promoting, and managing the Special Issue.

Stefania Gentili, Rita Di Giovambattista, Robert Shcherbakov, and Filippos Vallianatos
Editors

Editorial

Editorial of the Special Issue "Statistics and Pattern Recognition Applied to the Spatio-Temporal Properties of Seismicity"

Stefania Gentili [1,*], Rita Di Giovambattista [2], Robert Shcherbakov [3,4] and Filippos Vallianatos [5]

[1]   National Institute of Oceanography and Applied Geophysics—OGS, 33100 Udine, Italy
[2]   National Institute of Geophysics and Volcanology (INGV), 00143 Rome, Italy; rita.digiovambattista@ingv.it
[3]   Department of Earth Sciences, Western University, London, ON N6A 5B7, Canada; rshcherb@uwo.ca
[4]   Department of Physics and Astronomy, Western University, London, ON N6A 3K7, Canada
[5]   Department of Geophysics-Geothermics, Faculty of Geology and Geoenvironment, School of Sciences, National and Kapodistrian University of Athens, GR 15784 Athens, Greece; fvallian@geol.uoa.gr
*   Correspondence: sgentili@inogs.it

Citation: Gentili, S.; Di Giovambattista, R.; Shcherbakov, R.; Vallianatos, F. Editorial of the Special Issue "Statistics and Pattern Recognition Applied to the Spatio-Temporal Properties of Seismicity". *Appl. Sci.* **2022**, *12*, 4504. https://doi.org/10.3390/app12094504

Received: 15 April 2022
Accepted: 21 April 2022
Published: 29 April 2022

Publisher's Note: MDPI stays neutral with regard to jurisdictional claims in published maps and institutional affiliations.

1. Summary of the Special Issue Contents

Due to the significant increase in the availability of new data in recent years, as a result of the expansion of available seismic stations, laboratory experiments, and the availability of increasingly reliable synthetic catalogs, considerable progress has been made in understanding the spatiotemporal properties of earthquakes. The study of the preparatory phase of earthquakes and the analysis of past seismicity has led to the formulation of seismicity models for the forecasting of future earthquakes or to the development of seismic hazard maps. The results are tested and validated by increasingly accurate statistical methods. A relevant part of the development of many models is the correct identification of seismicity clusters and scaling laws of background seismicity. In this collection, we present eight innovative papers that address all the above topics.

The occurrence of strong earthquakes (mainshocks) is analyzed from different perspectives in this Special Issue.

Ref. [1] proposes analysis using a medium-term earthquake prediction method (EEPAS) applied to California and New Zealand and analyzes the trade-off between time and the area identified by precursor seismicity.

Ref. [2] aims to establish the mechanical stability of a fault system by analyzing modulations of seismic activity as a function of known perturbations, in order to assess how unstable the faults are for additional stress. The method is applied to Greek seismicity.

Ref. [3] proposes a pattern recognition approach to identify areas where strong earthquakes occur, for application in seismic hazard assessment studies. The method is applied to North and South America, Eurasia, and the Pacific Rim.

Three other papers are related to triggered and clustered seismicity analyses (foreshock and aftershock).

Ref. [4] presents the modeling of aftershock occurrence rates by comparing Omori-Utsu and ETAS laws, and estimates the probability of having the largest aftershock forecasted during a given future time interval using the extreme value theory and the Bayesian predictive framework. A retrospective forecasting of three sequences in Alaska is performed.

Ref. [5] describes a new cluster identification procedure, MAP-DBSCAN, and successfully compares its performance with that of other existing methods in the literature by using synthetic catalogs. The method is then applied to obtain a characterization of Greek seismicity.

Ref. [6] proposes the use of foreshock and aftershock data together with their mainshocks to improve an earthquake prediction technique based on spatially smoothed seismicity. The method is applied to a global catalog with two different magnitude thresholds, 5.5 and 6.5, showing improved performance.

The last two papers in the collection are closely related to the topic of the previous papers.

Ref. [7] shows an extended version of the maximum likelihood estimation method for estimating the parameters of the tapered Gutenberg–Richter distribution and their uncertainties, in the case of catalogs with a time-varying magnitude of completeness. The method is tested on synthetic catalogs and the global centroid moment tensor catalog.

Ref. [8] proposes an algorithm to simulate synthetic catalogs covering hundreds of thousands of years based on the ETAS model and seismogenic source data. The algorithm allows for obtaining a seismicity catalog, using the seismogenetic model of Italian seismicity derived from the DISS catalog, that reproduces sequences characterized by multiple mainshocks of similar magnitude, a typical aspect of northern and central Apennine seismicity.

2. Conclusions

The eight papers published in this collection represent a non-exhaustive list of the most recent leading topics in the fields of statistical seismology and pattern recognition applied to the spatiotemporal evolution of seismicity. Given the complexity of the topic, a rigorous methodology to the approach to the study of spatiotemporal properties of seismicity is needed. Descriptions of methods and implementations of modern and advanced methodologies from multidisciplinary approaches are needed for a shared understanding of the topic and as a starting point for new research. Therefore, this collection stands as an important starting point to outline the issues of interest as well as new challenges in this field.

Conflicts of Interest: The authors declare no conflict of interest.

References

1. Rastin, S.J.; Rhoades, D.A.; Christophersen, A. Space-Time Trade-Off of Precursory Seismicity in New Zealand and California Revealed by a Medium-Term Earthquake Forecasting Model. *Appl. Sci.* **2021**, *11*, 10215. [CrossRef]
2. Zaccagnino, D.; Telesca, L.; Doglioni, C. Different fault response to stress during the seismic cycle. *Appl. Sci.* **2021**, *11*, 9596. [CrossRef]
3. Dzeboev, B.A.; Gvishiani, A.D.; Agayan, S.M.; Belov, I.O.; Karapetyan, J.K.; Dzeranov, B.V.; Barykina, Y.V. System-Analytical Method of Earthquake-Prone Areas Recognition. *Appl. Sci.* **2021**, *11*, 7972. [CrossRef]
4. Sedghizadeh, M.; Shcherbakov, R. The Analysis of the Aftershock Sequences of the Recent Mainshocks in Alaska. *Appl. Sci.* **2022**, *12*, 1809. [CrossRef]
5. Bountzis, P.; Papadimitriou, E.; Tsaklidis, G. Identification and Temporal Characteristics of Earthquake Clusters in Selected Areas in Greece. *Appl. Sci.* **2022**, *12*, 1908. [CrossRef]
6. Taroni, M.; Akinci, A. A New Smoothed Seismicity Approach to Include Aftershocks and Foreshocks in Spatial Earthquake Forecasting: Application to the Global Mw $\geq$ 5.5 Seismicity. *Appl. Sci.* **2021**, *11*, 10899. [CrossRef]
7. Taroni, M.; Selva, J.; Zhuang, J. Estimation of the Tapered Gutenberg-Richter Distribution Parameters for Catalogs with Variable Completeness: An Application to the Atlantic Ridge Seismicity. *Appl. Sci.* **2021**, *11*, 12166. [CrossRef]
8. Console, R.; Vannoli, P.; Carluccio, R. Physics-Based Simulation of Sequences with Foreshocks, Aftershocks and Multiple Main Shocks in Italy. *Appl. Sci.* **2022**, *12*, 2062. [CrossRef]

Article

Space-Time Trade-Off of Precursory Seismicity in New Zealand and California Revealed by a Medium-Term Earthquake Forecasting Model

Sepideh J. Rastin *, David A. Rhoades and Annemarie Christophersen

GNS Science, Lower Hutt 5010, New Zealand; d.rhoades@gns.cri.nz (D.A.R.); a.christophersen@gns.cri.nz (A.C.)
* Correspondence: s.rastin@gns.cri.nz

Abstract: The 'Every Earthquake a Precursor According to Scale' (EEPAS) medium-term earthquake forecasting model is based on the precursory scale increase (Ψ) phenomenon and associated scaling relations, in which the precursor magnitude M_P is predictive of the mainshock magnitude M_m, precursor time T_P and precursor area A_P. In early studies of Ψ, a relatively low correlation between T_P and A_P suggested the possibility of a trade-off between time and area as a second-order effect. Here, we investigate the trade-off by means of the EEPAS model. Existing versions of EEPAS in New Zealand and California forecast target earthquakes of magnitudes $M > 4.95$ from input catalogues with $M > 2.95$. We systematically vary one parameter each from the EEPAS distributions for time and location, thereby varying the temporal and spatial scales of these distributions by two orders of magnitude. As one of these parameters is varied, the other is refitted to a 20-year period of each catalogue. The resulting curves of the temporal scaling factor against the spatial scaling factor are consistent with an even trade-off between time and area, given the limited temporal and spatial extent of the input catalogue. Hybrid models are formed by mixing several EEPAS models, with parameter sets chosen from points on the trade-off line. These are tested against the original fitted EEPAS models on a subsequent period of the New Zealand catalogue. The resulting information gains suggest that the space–time trade-off can be exploited to improve forecasting.

Keywords: earthquake forecasting; precursors; statistical seismology; earthquake likelihood models; seismicity patterns; New Zealand; California

Citation: Rastin, S.J.; Rhoades, D.A.; Christophersen, A. Space-Time Trade-Off of Precursory Seismicity in New Zealand and California Revealed by a Medium-Term Earthquake Forecasting Model. *Appl. Sci.* **2021**, *11*, 10215. https://doi.org/10.3390/app112110215

Academic Editor: Robert Shcherbakov

Received: 3 October 2021
Accepted: 20 October 2021
Published: 31 October 2021

Publisher's Note: MDPI stays neutral with regard to jurisdictional claims in published maps and institutional affiliations.

1. Introduction

Medium-term earthquake forecasting with time-varying models is becoming increasingly important for operational earthquake forecasting and the development of seismic hazard models. For example, the New Zealand medium-term forecast model has end users interested in time-varying earthquake hazards and risk, including the land use planning and building sector, central and local government agencies and the insurance industry [1–3].

Empirical observations of precursory seismicity patterns have an important role in aiding the development of earthquake forecasting models [4–10]. One such pattern is the precursory scale increase (Ψ) phenomenon, which is an increase in the magnitude and rate of occurrence of small earthquakes [11,12]. Individual examples of Ψ were identified by examining the seismicity in arbitrary frames of space and time preceding the occurrence of a major earthquake, such as in Figure 1 for the 2014 Napa, California earthquake. A magnitude versus time plot (Figure 1b) and a cumulative magnitude anomaly (Cumag) plot (Figure 1c) were used to identify the onset of precursory seismicity [12]. The onset is marked by the minimum of the Cumag plot. The precursor time T_P is then found as the time between the onset and origin time of the major earthquake. The space–time frame was chosen to informally maximize the increase in magnitude and seismicity rate at the time of the onset. Each example of Ψ provided a value of the mainshock magnitude M_m, precursory magnitude M_P, precursor time T_P and precursory area A_P (Figure 1c), within which the precursors, major earthquake and aftershocks all occurred.

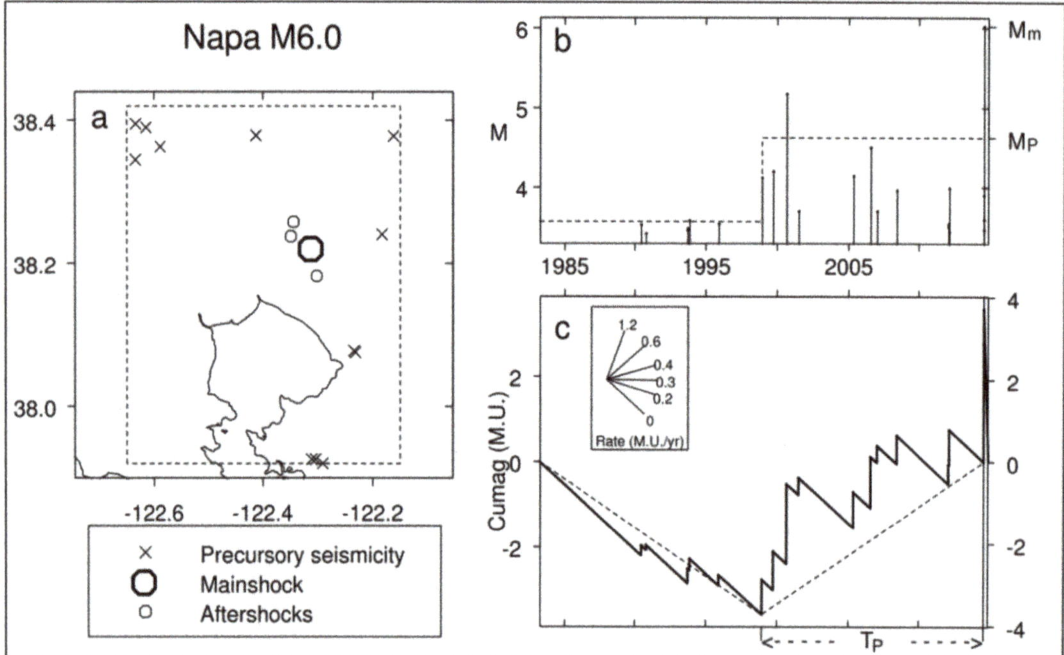

Figure 1. Identification of Ψ phenomenon for the August 2014 *M*6.0 South Napa, California earthquake. (**a**) The precursory area A_P (dashed rectangle) with the epicenters of the precursory seismicity, mainshocks and aftershocks. (**b**) Magnitude versus time of prior and precursory earthquakes. Dashed lines show the precursory increase in magnitude level. M_m is the main shock magnitude, and M_P is the precursor magnitude. (**c**) Changes in the cumulative magnitude anomaly (Cumag) over time; see [12] for the definition. Dashed lines show the precursory increase in the seismicity rate in 1998. The protractor translates the Cumag slope into the seismicity rate in magnitude units per year (M.U. yr^{-1}). T_P is the precursor time.

From the combined identifications of Ψ from four well-catalogued regions, it was found that M_m, M_P, T_P and A_P were all positively correlated [12]. In particular, three scaling relations (Figure 2) allowed M_m, T_P and A_P to be predicted from M_P, defined as the average magnitude of the three largest precursory earthquakes. These three predictive relations became the basis for the 'Every Earthquake a Precursor According to Scale' (EEPAS) medium-term earthquake forecasting model [13].

Although M_m, M_P, T_P and A_P were all positively correlated, A_P and T_P were less correlated than the other pairs of variables, as shown by the low value of the coefficient of determination R^2 in Figure 3a compared with those in Figure 2a–c. In Figure 2, we highlighted the earthquakes for which A_P was high and T_P was low or vice versa relative to the fitted relations, a condition that is not uncommon. The same earthquakes are highlighted in Figure 3. Remarkably, the product of T_P and A_P was highly correlated with M_m, as seen in Figure 3b, with R^2 being higher than any of those values in Figure 2. These features pointed to a trade-off between A_P and T_P. However, the origin of this trade-off was not clear. Could it have a physical origin related to, say, the tectonic setting or seismicity rate [14–16], or could it be a statistical side-effect? For example, in this case, if log T_P and log A_P were independently correlated with M_m, then their sum would be correlated even better, such as in Figure 3b.

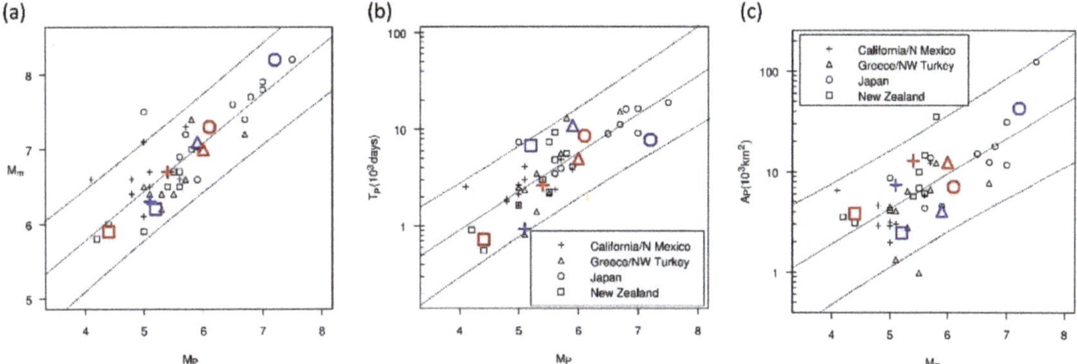

Figure 2. Predictive scaling relations and 95% tolerance limits derived from 47 examples of ψ from four regional earthquake catalogues, taken after [12]. (**a**) Mainshock magnitude M_m versus precursor magnitude M_P (coefficient of determination $R^2 = 71\%$). (**b**) Precursor time T_P versus M_P ($R^2 = 65\%$). (**c**) Precursory area A_P versus M_P ($R^2 = 48\%$). Enlarged and colored points are for 1990 Weber (blue square), 1968 Puysegur Bank (red square), 1969 E. Hokkaido (blue circle), 2000 W. Tottori (red circle), 1948 Karpathos (blue triangle), 1983 Kefallonia (red triangle), 1966 Colorado D. (blue cross) and 1980 S. Cascadia (red cross).

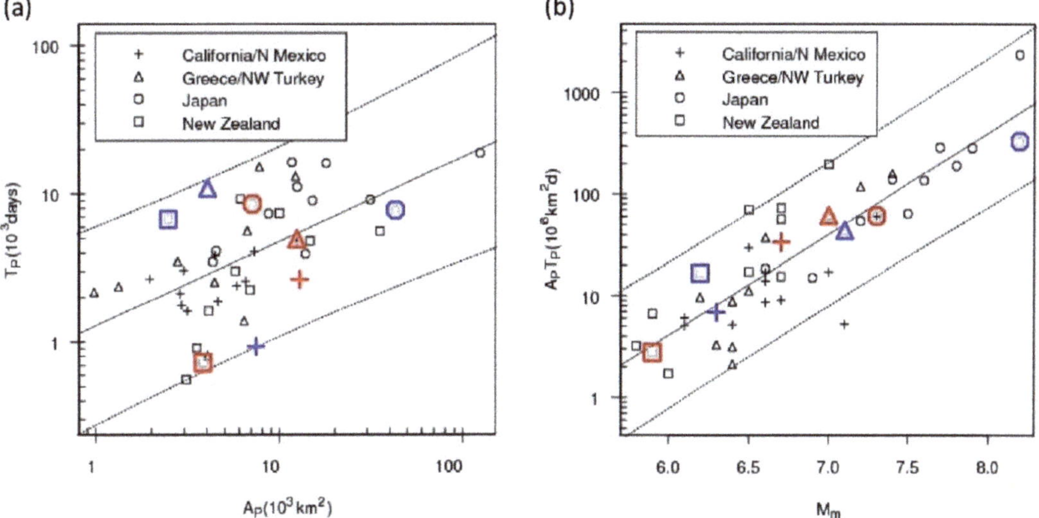

Figure 3. Scaling relations and 95% tolerance limits derived from 47 examples of Ψ from four regional earthquake catalogues, taken after [12]. (**a**) Precursor time T_P versus precursory area A_P ($R^2 = 34\%$). (**b**) Product of A_P and T_P versus mainshock magnitude M_m ($R^2 = 75\%$). Symbols are enlarged and colored as in Figure 2.

A study of the Ψ phenomenon in synthetic earthquake catalogues shed new light on the matter [17]. It was found that, in a synthetic catalogue generated by the earthquake simulator RSQSim [18,19], two or more equally plausible identifications of Ψ could be found for individual mainshocks. These identifications presented very different T_P and A_P values, consistent with a hypothetical space–time trade-off.

The evidence for the trade-off, whatever its origin, can also be strengthened through applications of the EEPAS model. One example was the EEPAS model fitted with different fixed lead times [20]. The lead time is defined as the time interval between the start of the catalogue and the origin time of a target earthquake. It was found that as the lead time increases, the mean of the EEPAS time distribution increases, and the variance of the location distribution decreases. The time and spatial scales involved varied by about a factor of two. Here, we aim to further understand the space–time trade-off by fitting the EEPAS time distribution with a fixed spatial distribution and the spatial distribution with a fixed time distribution.

In the next section, we review the defining equations of the EEPAS model and then describe the method and data for the present study. Our results show how the space–time trade-off is revealed through constrained fitting of the EEPAS model to the New Zealand and California catalogues. Finally, we indicate by way of a simple New Zealand example how the space–time trade-off might be exploited for improving the performance of medium-term earthquake forecasts.

2. EEPAS Forecasting Model

Although inspired by the Ψ predictive scaling relations (Figure 2), the EEPAS model does not involve the identification of precursory seismicity for individual major earthquakes. It treats every earthquake as a potential precursor of future larger earthquakes to follow in the medium term [13]. Depending on the magnitude, this period can range from months to decades. The model has a background component and a time-varying component. The background component is a smoothed seismicity model, with the spatial distribution depending on the proximity to the location of past earthquakes. It is, in principle, time-invariant, but it is updated at the origin time of each contributing earthquake. The time-varying component, based on the Ψ predictive relations, is obtained by summing the contributions from all past earthquakes after a starting time t_0 and exceeding an input magnitude threshold m_0. The expected earthquake occurrence rate density is a function of the time, magnitude and location denoted by λ. For times $t > t_0$, magnitudes m exceeding a target threshold m_c and locations (x,y) within a region of surveillance R, the total rate density takes the following form:

$$\lambda(t, m, x, y) = \mu \lambda_0(t, m, x, y) + \sum_{t_i \geq t_0, m_i \geq m_0} \eta(m_i) \lambda_i(t, m, x, y) \tag{1}$$

where μ is an adjustable mixing parameter representing the proportion of the forecast contributed by the background model component; λ_0 is the rate density of the background model; η is a normalizing function and t_i and m_i are the origin time and magnitude of the ith contributing earthquake, respectively. The contributing earthquakes come from a larger search region, which needs to be big enough to include all earthquakes that might affect the rate density within R. The contribution from the ith earthquake to the rate density is given by

$$\lambda_i(t, m, x, y) = w_i f(t|t_i, m_i) g(m|m_i) h(x, y|x_i, y_i, m_i), \tag{2}$$

in which w_i is a weighting factor and f, g and h are the densities of probability distributions which are based on the Ψ predictive scaling relations (Figure 2). These distributions depend on the magnitude m_i of the contributing earthquake. Following the notation in [20], the magnitude density g is a normal density of the following form:

$$g(m|m_i) = \frac{1}{\sigma_M \sqrt{2\pi}} \exp\left[-\frac{1}{2}\left(\frac{m - a_M - b_M m_i}{\sigma_M}\right)^2\right], \tag{3}$$

in which a_M, b_M and σ_M are adjustable parameters. The time density f is a lognormal density of the following form:

$$f(t|t_i, m_i) = \frac{H(t - t_i)}{(t - t_i)\sigma_T \ln(10)\sqrt{2\pi}} \exp\left[-\frac{1}{2}\left(\frac{\log(t - t_i) - a_T - b_T m_i}{\sigma_T}\right)^2\right], \tag{4}$$

in which $H(s) = 1$ if $s > 0$ and is 0 otherwise and a_T, b_T and σ_T are adjustable parameters. If all other parameters are fixed, the mean of the time distribution is proportional to 10^{a_T}.

Therefore, 10^{a_T} can be regarded as a temporal scaling factor. The location density h is a bivariate normal density of the following form:

$$h(x, y | x_i, y_i, m_i) = \frac{1}{2\pi\sigma_A^2 10^{b_A m_i}} \exp\left[-\frac{(x - x_i)^2 + (y - y_i)^2}{2\sigma_A^2 10^{b_A m_i}}\right],\tag{5}$$

in which σ_A and b_A are adjustable parameters. If all other parameters are fixed, the area occupied by the location distribution is proportional to σ_A^2. Therefore, σ_A^2 can be regarded as a spatial scaling factor.

The adjustable parameters are fitted to maximize the log likelihood of the target earthquakes in the region of surveillance over a fitting period (t_s, t_f) and a magnitude range (m_c, m_{max}). If the target earthquakes have coordinates $\left[(t_{ij}, m_{ij}, x_{ij}, y_{ij}), j = 1, \cdots, N\right]$, the space–time point process log likelihood [21,22] is given by

$$\ln L = \sum_{j=1}^{N} \ln \lambda(t_{ij}, m_{ij}, x_{ij}, y_{ij}) + \iint_R \int_{m_c}^{m_{max}} \int_{t_s}^{t_f} \lambda(t, m, x, y) dt dm dx dy.\tag{6}$$

Information gain statistics compare the performance of different models with the same data [23]. For models with the same number of fitted parameters or for testing pre-fitted models on an independent data set, the information gain per earthquake $I(X, Y)$ of one model X over another model Y is given by

$$I(X, Y) = (\ln L_X - \ln L_Y)/N.\tag{7}$$

where $\ln L_X$ is the log-likelihood of model X and N is the number of target earthquakes [24].

3. Method

The EEPAS model is usually fitted with a time lag to prevent any influence on the parameters from short-term clustering. Here, a time lag of 50 days was applied. This means that no precursory earthquake contributed to the time-varying rate density until 50 days after its occurrence.

Two different weighting strategies are commonly adopted in applications of the EEPAS model: equal weighting and down-weighting of aftershocks. For down-weighting of aftershocks, the weight assigned to each earthquake depends on the ratio of the rate density of the background model to the rate density of an epidemic-type aftershock model at the time, magnitude and location of its occurrence. For details, see [13]. The down-weighted aftershocks strategy is preferable for investigating the space–time trade-off because it better respects the hierarchical nature of seismicity, as seen in aftershock occurrence as well as precursory seismicity [25–27].

We considered two versions of the down-weighted aftershocks EEPAS model, which we labeled EEPAS_1F. The models were called NZ EEPAS_1F and California EEPAS_1F. The model parameters are listed in Table 1. The values were slightly different from the models previously tested in the New Zealand and California testing centers of the Collaboratory for the Study of Earthquake Predictability (CSEP) since 2008 and 2006, respectively [28–30]. The differences were due to looser constraints imposed in the fitting of μ.

The surveillance and search regions for New Zealand and California are shown in Figures 4 and 5, respectively. Figures 4a and 5a show the locations of earthquakes with magnitudes M > 2.95 contributing to their fitting between times t_0 and t_f. Time t_0 is the beginning of 1951 for New Zealand and 1932 for California, while t_f is the end of 2006 for New Zealand and 2005 for California. Figures 4b and 5b show the locations of the target earthquakes with M > 4.95 between times t_s and t_f, where t_s is the beginning of 1987 for New Zealand and 1986 for California.

Table 1. EEPAS_1F model parameters for New Zealand (NZ) and California.

Parameter	Details	NZ EEPAS-1F	California EEPAS-1F
m_0	Minimum precursor magnitude	2.95 *	2.95 *
m_c	Minimum target magnitude	4.95 *	4.95 *
m_u	Maximum target magnitude	10.05 *,!	10.05 *!
b_{GR}	Gutenberg–Richter b-value	1.16 †	1.0 †
a_M	Equation (3)	1.10 †	1.74 †
b_M	Equation (3)	1.0 *	1.0 *
σ_M	Equation (3)	0.35 †	0.60 †
a_T	Equation (4)	1.44 †	2.11 †
b_T	Equation (4)	0.43 †	0.40 †
σ_T	Equation (4)	0.53 †	0.43 †
b_A	Equation (5)	0.37 †	0.35 †
σ_A	Equation (5)	1.16 †	0.88 †
μ	Equation (8)	0.18 †	0.27 †

* Fixed. † Fitted. ! Standard threshold used for CSEP models.

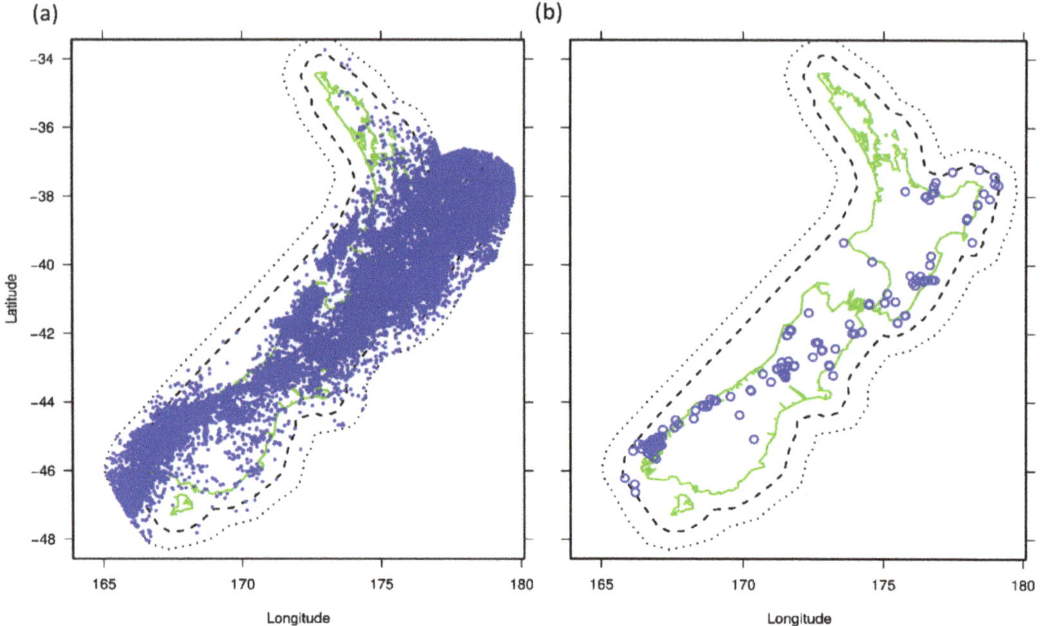

Figure 4. Maps of New Zealand seismicity, including the region of surveillance (inner dashed polygon), the search region (outer dotted polygon) and locations of earthquakes with magnitudes (**a**) M > 2.95 with a hypocentral depth ≤45 km from 1951 to 2006 and (**b**) M > 4.95 with a hypocentral depth ≤40 km from 1987 to 2006 in the region of surveillance (158 target earthquakes).

To investigate the space–time trade-off, we varied the EEPAS model parameters in a controlled way. Starting with the parameter sets listed in Table 1, we separately changed the EEPAS_1F parameters σ_A and a_T while the other parameters, except the mixing parameter μ, remained fixed at their previously fitted values. We changed σ_A in seven steps in either direction away from its optimal value (Table 2) and obtained the corresponding values of the temporal scaling factor σ_A^2. Subsequently, we changed the a_T values in a similar manner (Table 3) and obtained the corresponding values of the temporal scaling factor 10^{a_T}. Over seven steps, each of the controlled scaling factors varied by an order of magnitude on either side of the optimal fit. For each controlled value of a_T or σ_A, two free parameters, μ and either σ_A or a_T, were refitted to maximize the likelihood of target earthquakes in the region of surveillance over time period (t_s, t_f).

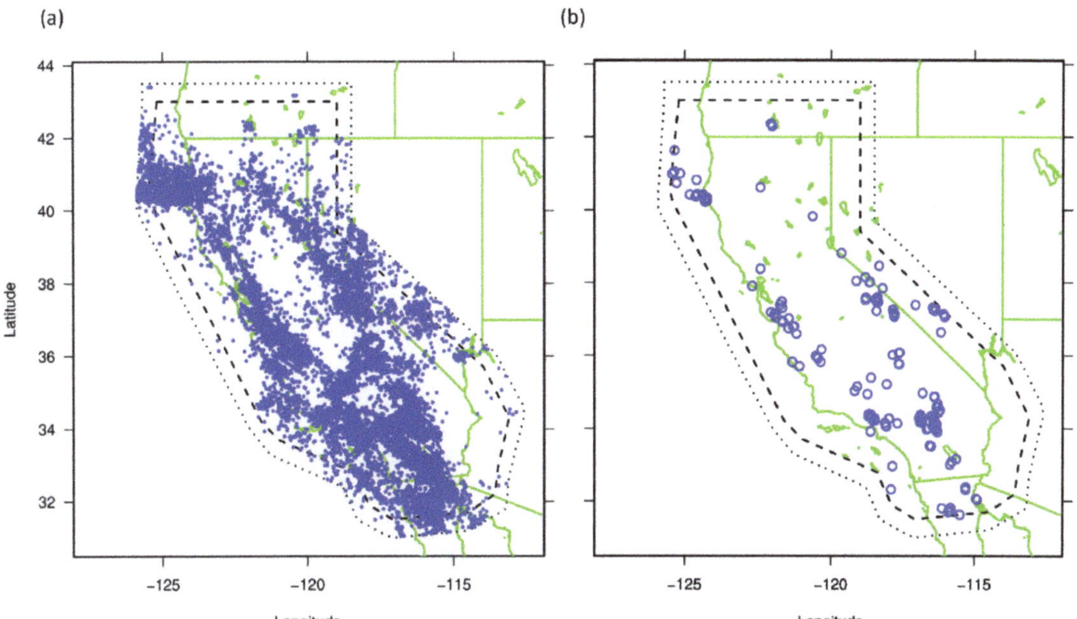

Figure 5. Maps of California's seismicity, including the region of surveillance (inner dashed polygon), search region (outer dotted polygon), and locations of earthquakes with magnitudes (**a**) M > 2.95 and hypocentral depths ≤30 km from 1932 to 2004 and (**b**) M > 4.95 and hypocentral depths ≤30 km from 1986 to 2005 in the region of surveillance (155 target earthquakes).

Table 2. Controlled values of σ_A in EEPAS_1F model for New Zealand (NZ) and California.

NZ EEPAS-1F	California EEPAS-1F
0.34	0.26
0.41	0.31
0.49	0.37
0.58	0.44
0.69	0.53
0.82	0.97
0.97	0.74
1.16 [†]	0.88 [†]
1.38	1.05
1.64	1.25
1.95	1.49
2.32	1.77
2.75	2.10
3.27	2.50
3.89	2.98

[†] Fitted.

Table 3. Controlled values of a_T in EEPAS_1F models for NZ and California.

NZ EEPAS-1F	California EEPAS-1F
2.49	3.16
2.34	3.01
2.19	2.86
2.04	2.71
1.89	2.56
1.74	2.41
1.59	2.26
1.44 [†]	2.11 [†]
1.29	1.96
1.14	1.81
0.99	1.66
0.84	1.51
0.69	1.36
0.54	1.21
0.39	1.06

[†] Fitted.

4. Results

The likelihood of the refitted models declined with each step change in the controlled parameter away from its optimal value, as shown for New Zealand in Figure 6. The results for California were similar. The log-likelihood of the refitted model is plotted against the controlled spatial scaling factor in Figure 6a and against the temporal scaling factor in Figure 6b. An order of magnitude change in each scaling factor induced a modest reduction in the log-likelihood. The maximum reduction of about 34 units corresponded to an information loss per earthquake of about 0.2 relative to the overall optimal fit.

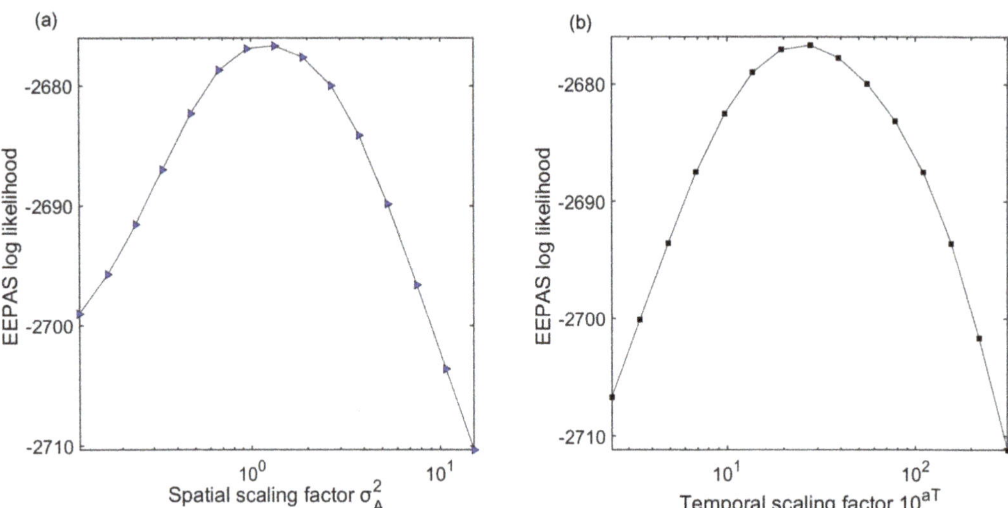

Figure 6. Log-likelihood of EEPAS model fitted with controlled values of (**a**) σ_A (Table 2) and (**b**) a_T (Table 3) to the New Zealand earthquake catalogue.

The refitted mixing parameter μ tended to increase as the controlled parameter shifted further away from its optimal value, as shown for New Zealand in Figure 7. Again, the results were similar for California. The variation of μ with the spatial scaling factor is shown in Figure 7a and against the temporal scaling factor in Figure 7b. The values of μ increased from about 0.15 at the optimal fit to greater than 0.5 when the temporal or spatial scaling factors were changed by an order of magnitude. The μ value represents the proportional contribution of the background model to the total EEPAS model rate density. Higher μ values thus indicate a greater contribution of the background component and a smaller contribution of the time-varying component. In other words, higher μ values indicate that there were fewer target earthquakes with precursors matching the changed spatial and temporal distributions.

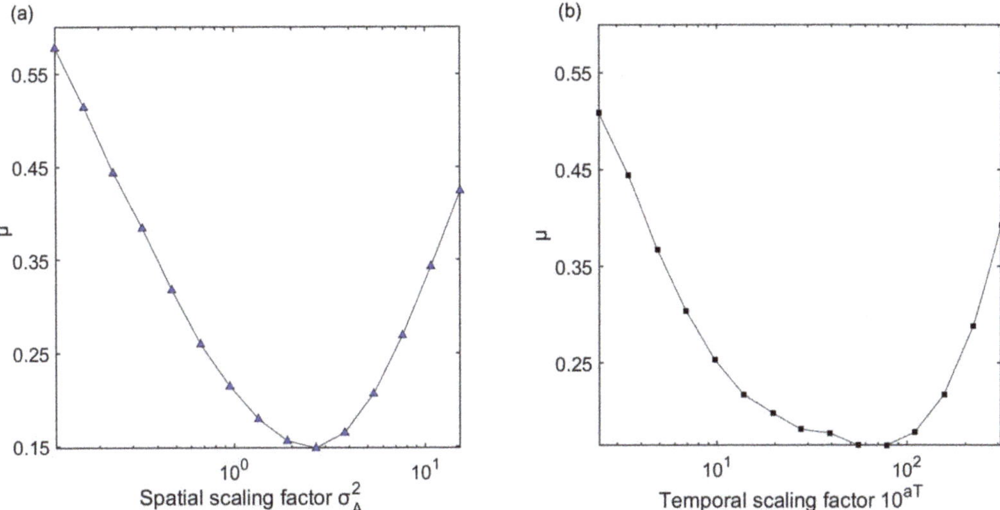

Figure 7. Fitted values of mixing parameter μ ($0 \leq \mu \leq 1$) of the EEPAS model fitted with controlled values of (**a**) σ_A and (**b**) a_T to the New Zealand earthquake catalogue.

As the controlled parameter was changed, the refitted values of the other parameters changed in a way that was consistent with the notion of a space–time trade-off. The results are shown for New Zealand in Figure 8a and for California in Figure 8b.

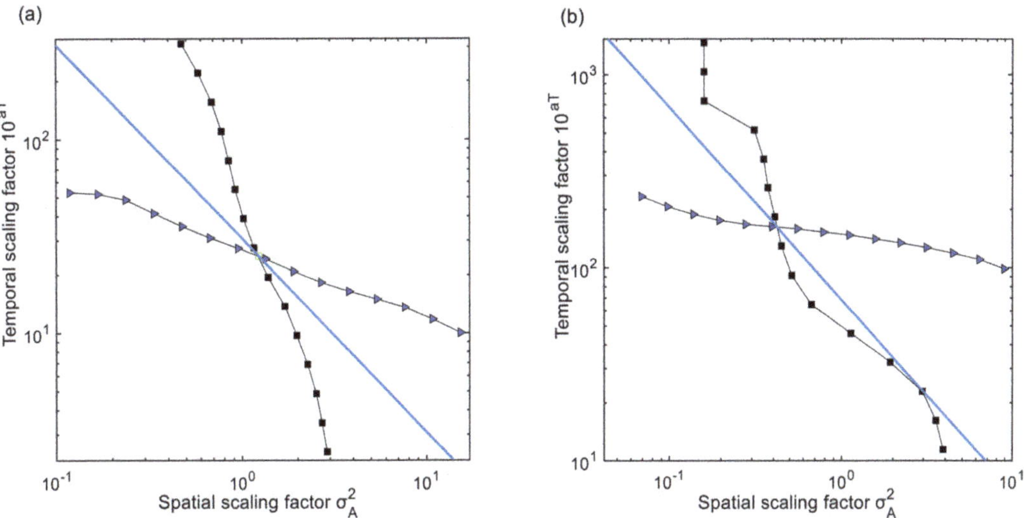

Figure 8. Trade-off of spatial and temporal scaling factors $\sigma_A{}^2$ and 10^{a_T}, respectively, revealed by the fit of the EEPAS model with controlled values of σ_A (blue triangles) and a_T (black squares). The straight line with a slope of -1 represents an even trade-off between space and time. (**a**) New Zealand. (**b**) California.

In each plot, the pairs of scaling factors resulting from controlling σ_A are shown as blue triangles, and those resulting from controlling a_T are shown as black squares. The temporal scaling factor decreased as the controlled spatial scaling factor increased, and the spatial scaling factor decreased as the controlled temporal scaling factor increased. However, the curves had different slopes depending on whether σ_A or a_T was the controlled variable. An even trade-off line with a slope of -1 is drawn through the intersection of the two curves (straight blue line in Figure 8a,b). Its slope lies between the average slopes of the two controlled fitting curves.

5. Discussion

As seen in Figure 8, the controlled fits produced two curves which did not lie on the even trade-off line but instead had higher or lower slopes. This result can be explained by the limitations on the length of the catalogue and the size of the search region. The fitted parameters could only adjust to the precursors that were contained in the catalogue and not to those that were screened out by such limitations. We now consider in detail the trend of the fitted σ_A value away from the even trade-off line for the controlled values of a_T. The trend of the other curve can be explained similarly.

As a_T was stepped down to lower values (i.e., the time scale was shortened), fitting the trade-off required earthquakes at increasingly longer distances from the target earthquakes. However, at longer distances, more precursory events were screened out by the spatial limitation on the input catalogue. The precursors of the largest earthquakes in the target magnitude range would be most affected by the spatial limitation because they had larger precursory areas (Figure 2). The spatial limitation at small a_T values forced the fitted values of σ_A to increasingly fall below the even trade-off line. On the other hand, as a_T was stepped up to higher values, the precursory time scale became longer and exceeded the available lead time. This temporal limitation most affected the largest earthquakes in the target magnitude range, which had the longest precursor times (Figure 2). Thus, more and more precursory earthquakes on the specified time scale were screened out by the limited time span of the catalogue. The remaining precursors for fitting σ_A would be those at the lower end of the time distribution. Because of the space–time trade-off, these remaining precursors tended to be at longer distances than the screened-out events. This forced the fitted σ_A to increasingly exceed the even trade-off line.

The space–time trade-off in the EEPAS model shows that as the mean of $f(t|m)$ in Equation (4) increased, the area of the fitted $h(x,y|m)$ in Equation (5) decreased and vice versa. This phenomenon can be interpreted in terms of the predictive scaling relations on which the EEPAS model is based (Figure 2). Figure 2 shows that T_P and A_P both increased with the precursory earthquake magnitude M_P. Similarly, in the EEPAS model, the mean of the time distribution $f(t|m)$ and the area of the location distribution $h(x,y|m)$ both increased with m. Now, the space–time trade-off observed in the EEPAS model can be interpreted in terms of the space–time distribution of precursors to an individual major earthquake; that is, the earliest precursors tend to occur very close to the source, and the later precursors to occupy a wide area around the source. This interpretation only applies to precursors occurring more than 50 days before the mainshock because of the time lag applied for EEPAS model fitting here.

The existence of this trade-off raises the question of how it can be exploited to improve the performance of the EEPAS model. The EEPAS model treats the time and location as independent variables, but the trade-off implies that they are correlated. We will illustrate how to improve forecasting by forming hybrid models. The hybrid models are mixtures of three EEPAS models with the values of a_T and σ_A chosen from points on the even trade-off line with a slope of -1. We constructed two models, Hybrid_1F and Hybrid_1R, starting from two different EEPAS models: EEPAS_1F and EEPAS_1R, respectively. EEPAS_1R was similar to EEPAS_1F in nearly all aspects, apart from having fewer optimized parameters. Its fixed and optimized parameters are given in Table 4. An important difference between the two models was that EEPAS_1F (Table 1) had a larger value of σ_T than EEPAS_1R

(Table 4). The parameter σ_T was optimized in the fitting of EEPAS_1F but not in EEPAS_1R. In prospective testing over 10 years in the New Zealand CSEP testing center, EEPAS_1F significantly outperformed EEPAS_1R [29].

Table 4. EEPAS_1R model parameters for New Zealand.

Parameter	Value
m_0	2.95 *
m_c	4.95 *
m_u	10.05 *
b_{GR}	1.16 †
a_M	1.00 †
b_M	1.0 *
σ_M	0.32 *
a_T	1.40 †
b_T	0.40 *
σ_T	0.23 *
b_A	0.35 *
σ_A	1.74 †
μ	0.24 †

* Fixed. † Fitted.

To construct the hybrid models, we replaced the time-varying component of each model's rate density with the average rate density of the three models, with the values of a_T and σ_A chosen from the trade-off line. The three models were the original one and two others formed by an arbitrary increase and decrease in a_T of $\Delta = 0.5$. For an increase in Δ in a_T, the corresponding value of σ_A on the trade-off line was found by multiplying the original σ_A by $10^{-0.5\Delta}$. The other parameters, including μ and σ_T, remained unchanged at their values in Tables 1 and 4. Using information gain statistics, we compared the performance of the EEPAS_1F, EEPAS_0F, Hybrid_1F and Hybrid_1R models. For this, we used a test period from 2007 to 2017, during which there were 259 target earthquakes with magnitudes M > 4.95. Hybrid_1R outperformed all the other models, and EEPAS_1R was the weakest model (Figure 9). Figure 9a shows the information gain of EEPAS_1F, and Figure 9b shows that of Hybrid_1R over the other models. Both hybrid models and EEPAS_1F outperformed EEPAS_1R with 95% confidence according to the *T*-test [24].

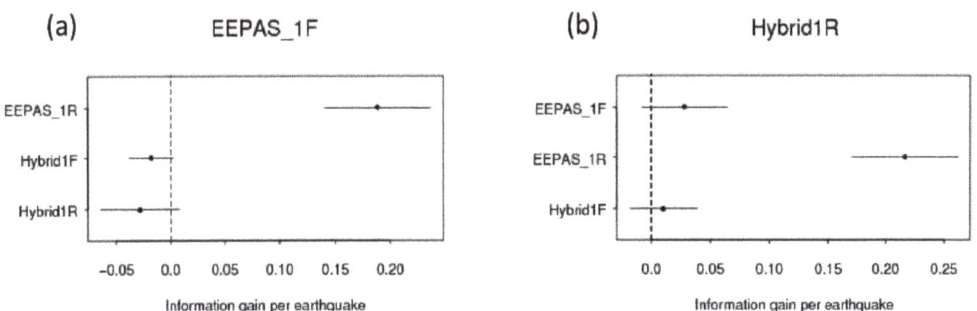

Figure 9. Information gain per earthquake and 95% confidence interval of the (**a**) EEPAS_1F model and (**b**) Hybrid_1R model compared with other models during the test period of 2007–2017 in the New Zealand testing region (259 target earthquakes with M > 4.95).

This simple example of hybrid formation, even without fitting additional parameters, suggests that it might be possible to use the space–time trade-off to improve forecasting. However, much more work needs to be done to construct a formal method for optimal inclusion of the trade-off in the fitting of the EEPAS model. The temporal and spatial limitations of the catalogue are clearly among the issues to be considered. The spatial limitations can be resolved if a global catalogue is used, but then a higher threshold magnitude of completeness would apply. That in turn imposes further limitations. Additionally, there is evidence that the precursor time distribution is dependent on the strain rate in the vicinity of a target earthquake [17]. This dependence would have to be included in a global model. Temporal limitations can also be partly resolved by introducing a fixed lead time for all target earthquakes and then compensating for the lead time using the method described in [20].

6. Conclusions

A space–time trade-off of precursory seismicity has been investigated by repeated refitting of the EEPAS earthquake forecasting model to the catalogues of New Zealand and California. In a sequence of controlled fits, the temporal scaling parameter was constrained to vary in steps ranging over two orders of magnitude with the spatial scaling parameter before being refitted, and vice versa. The two resulting curves of the temporal scaling factor against the spatial scaling factor differed depending on which parameter was controlled and which was fitted. However, both curves were consistent with an even trade-off between space and time once the temporal and spatial limits of the contributing earthquake data were considered. As the controlled parameter deviated further from its optimal value, the likelihood of the refitted model decreased. In addition, the refitted model had an increasingly large background component and a diminishing time-varying component.

The trade-off implies that the earliest precursors to a major earthquake tend to occur very close to its source and that the later precursors occupy a wide area around the source. A simple example in which hybrid forecasts were created by mixing several EEPAS models with parameters chosen from the trade-off line suggests that it should be possible to exploit the trade-off for improved forecasting. However, more research is needed to develop a formal method for routinely incorporating the space–time trade-off into medium-term earthquake forecasts.

Author Contributions: Conceptualization, S.J.R., D.A.R. and A.C.; methodology, D.A.R.; software, S.J.R. and D.A.R.; formal analysis, S.J.R. and D.A.R.; writing—original draft preparation, S.J.R. and D.A.R.; writing—review and editing, S.J.R., D.A.R. and A.C. All authors have read and agreed to the published version of the manuscript.

Funding: This research was funded by the Strategic Science Investment Fund (SSIF) of the Ministry of Business, Innovation and Employment of New Zealand.

Institutional Review Board Statement: Not applicable.

Informed Consent Statement: Not applicable.

Data Availability Statement: The New Zealand Earthquake Catalogue was obtained from GeoNet. Available online: http://www.geonet.org.nz (accessed 30 September 2021). Earthquake data for the California region came from the Advanced National Seismic System (ANSS) Worldwide Earthquake Catalog, which is contributed to by members of the U.S. Council of the National Seismic System and maintained by the Northern California Earthquake Data Center. Available online: www.ncedc.org/anss/catalog-search.html (accessed 30 September 2021).

Acknowledgments: We acknowledge the New Zealand GeoNet project and its sponsors EQC, GNS Science, LINZ, NEMA and MBIE for providing the data used in this study. We would like to thank Stephen Bannister and Rob Buxton for their constructive internal reviews of an earlier version of this manuscript. We also thank two anonymous reviewers. However, the authors are solely responsible for the final manuscript.

Conflicts of Interest: The authors declare no conflict of interest. The funders had no role in the design of the study; in the collection, analyses, or interpretation of data; in the writing of the manuscript, or in the decision to publish the results.

References

1. Gerstenberger, M.; McVerry, G.; Rhoades, D.; Stirling, M. Seismic Hazard Modeling for the Recovery of Christchurch. *Earthq. Spectra* **2014**, *30*, 17–29. [CrossRef]
2. Gerstenberger, M.C.; Rhoades, D.A.; McVerry, G.H. A Hybrid Time-Dependent Probabilistic Seismic-Hazard Model for Canterbury, New Zealand. *Seismol. Res. Lett.* **2016**, *87*, 1311–1318. [CrossRef]
3. Jordan, T.; Chen, Y.; Gasparini, P.; Madariaga, R.; Main, I.; Marzocchi, W.; Papadopoulos, G.; Sobolev, G.; Yamaoka, K.; Zschau, J. Operational Earthquake Forecasting: State of Knowledge and Guidelines for Utilization. *Ann. Geophys.* **2011**, *54*, 316–391.
4. Rikitake, T. Earthquake Prediction and Warning. *Interdiscip. Sci. Rev.* **1978**, *3*, 58–70. [CrossRef]
5. UNESCO; International Symposium on Earthquake Prediction. Earthquake prediction. In *Proceedings of the International Symposium on Earthquake Prediction, Tokyo, Japan, 1984*; UNESCO: Paris, France.
6. Kossobokov, V.G.; Maeda, K.; Uyeda, S. Precursory Activation of Seismicity in Advance of the Kobe, 1995, M = 7.2 Earthquake. *Pure Appl. Geophys.* **1999**, *155*, 409–423. [CrossRef]
7. Gulia, L.; Wiemer, S. Real-time discrimination of earthquake foreshocks and aftershocks. *Nature* **2019**, *574*, 193–199. [CrossRef]
8. Zavyalov, A. Medium-term prediction of earthquakes from a set of criteria: Principles, methods, and implementation. *Russ. J. Earth Sci.* **2005**, *7*, 51–73. [CrossRef]
9. Holliday, J.R.; Chen, C.-C.; Tiampo, K.F.; Rundle, J.B.; Turcotte, D.L.; Donnellan, A. A RELM Earthquake Forecast Based on Pattern Informatics. *Seismol. Res. Lett.* **2007**, *78*, 87–93. [CrossRef]
10. Chorozoglou, D.; Iliopoulos, A.; Kourouklas, C.; Mangira, O.; Papadimitriou, E. Earthquake Networks as a Tool for Seismicity Investigation: A Review. *Pure Appl. Geophys.* **2019**, *176*, 4649–4660. [CrossRef]
11. Evison, F.; Rhoades, D. Precursory scale increase and long-term seismogenesis in California and Northern Mexico. *Ann. Geophys.* **2002**, *45*, 479–495.
12. Evison, F.F.; Rhoades, D.A. Demarcation and Scaling of Long-term Seismogenesis. *Pure Appl. Geophys.* **2004**, *161*, 21–45. [CrossRef]
13. Rhoades, D.A.; Evison, F.F. Long-range Earthquake Forecasting with Every Earthquake a Precursor According to Scale. *Pure Appl. Geophys.* **2004**, *161*, 47–72. [CrossRef]
14. Aki, K. A new view of earthquake and volcano precursors. *Earth Planets Space* **2004**, *56*, 689–713. [CrossRef]
15. Rhoades, D.A. Long-range earthquake forecasting allowing for aftershocks. *Geophys. J. Int.* **2009**, *178*, 244–256. [CrossRef]
16. Rhoades, D.A. Lessons and Questions from Thirty Years of Testing the Precursory Swarm Hypothesis. *Pure Appl. Geophys.* **2010**, *167*, 629–644. [CrossRef]
17. Christophersen, A.; Rhoades, D.A.; Colella, H.V. Precursory seismicity in regions of low strain rate: Insights from a physics-based earthquake simulator. *Geophys. J. Int.* **2017**, *209*, 1513. [CrossRef]
18. Richards-Dinger, K.; Dieterich, J.H. RSQSim Earthquake Simulator. *Seismol. Res. Lett.* **2012**, *83*, 983–990. [CrossRef]
19. Shaw, B.E.; Milner, K.R.; Field, E.H.; Richards-Dinger, K.; Gilchrist, J.J.; Dieterich, J.H.; Jordan, T.H. A physics-based earthquake simulator replicates seismic hazard statistics across California. *Sci. Adv.* **2018**, *4*, eaau0688. [CrossRef]
20. Rhoades, D.A.; Rastin, S.J.; Christophersen, A. The Effect of Catalogue Lead Time on Medium-Term Earthquake Forecasting with Application to New Zealand Data. *Entropy* **2020**, *22*, 1264. [CrossRef] [PubMed]
21. Daley, D.J.; Vere-Jones, D. *An Introduction to the Theory of Point Processes*; Springer: Berlin/Heidelberg, Germany; Volume II, 2008.
22. Ogata, Y.; Zhuang, J. Space–time ETAS models and an improved extension. *Tectonophysics* **2006**, *413*, 13–23. [CrossRef]
23. Vere-Jones, D. Probabilities and information gain for earthquake forecasting. In *Selected Papers From Volume 30 of Vychislitel'naya Seysmologiya*; 2003; pp. 104–114.
24. Rhoades, D.A.; Schorlemmer, D.; Gerstenberger, M.C.; Christophersen, A.; Zechar, J.D.; Imoto, M. Efficient testing of earthquake forecasting models. *Acta Geophys.* **2011**, *59*, 728–747. [CrossRef]
25. Evison, F.; Rhoades, D. Long-term seismogenesis and self-organized criticality. *Earth Planets Space* **2004**, *56*, 749–760. [CrossRef]
26. Huang, Y.; Saleur, H.; Sammis, C.; Sornette, D. Precursors, aftershocks, criticality and self-organized criticality. *Europhys. Lett.* **1998**, *41*, 43–48. [CrossRef]
27. Kossobokov, V.G. Earthquake prediction: Basics, achievements, perspectives. *Acta Geod. Geophys. Hungarica* **2004**, *39*, 205–221. [CrossRef]
28. Schneider, M.R.; Clements, R.; Rhoades, D.; Schorlemmer, D. Likelihood- and residual-based evaluation of medium-term earthquake forecast models for California. *Geophys. J. Int.* **2014**, *198*, 1307–1318. [CrossRef]
29. Rhoades, D.A.; Christophersen, A.; Gerstenberger, M.C.; Liukis, M.; Silva, F.; Marzocchi, W.; Werner, M.J.; Jordan, T.H. Highlights from the First Ten Years of the New Zealand Earthquake Forecast Testing Center. *Seismol. Res. Lett.* **2018**, *89*, 1229–1237. [CrossRef]
30. Zechar, J.D.; Schorlemmer, D.; Liukis, M.; Yu, J.; Euchner, F.; Maechling, P.J.; Jordan, T.H. The Collaboratory for the Study of Earthquake Predictability perspective on computational earthquake science. *Concurr. Comput. Pract. Exp.* **2010**, *22*, 1836–1847. [CrossRef]

applied
sciences

Article

Different Fault Response to Stress during the Seismic Cycle

Davide Zaccagnino [1,*], Luciano Telesca [2] and Carlo Doglioni [3,4]

1   Department of Physics, Sapienza University of Rome, 00185 Rome, Italy
2   Istituto di Metodologie per l' Analisi Ambientale (CNR-IMAA), 85050 Tito Scalo, Italy;
    luciano.telesca@imaa.cnr.it
3   Department of Earth Science, Sapienza University of Rome, 00185 Rome, Italy; carlo.doglioni@uniroma1.it
4   Istituto Nazionale di Geofisica e Vulcanologia (INGV), 00143 Rome, Italy
*   Correspondence: zaccagnino.1748720@studenti.uniroma1.it

Featured Application: This article introduces a method to establish the state of mechanical stability of a fault system by analyzing modulations of seismic activity as a function of known perturbations, i.e., tidal stress. In addition to providing useful information about the physics of fault systems, our method can be applied to evaluate how unstable faults are with respect to additional stress, and therefore forecast their future slip. *Mutatis mutandis*, our approach can also be adopted in other fields where it is of paramount interest to assess the loading state of a physical system alternating stability to sudden breaking.

Abstract: Seismic prediction was considered impossible, however, there are no reasons in theoretical physics that explicitly prevent this possibility. Therefore, it is quite likely that prediction is made stubbornly complicated by practical difficulties such as the quality of catalogs and data analysis. Earthquakes are sometimes forewarned by precursors, and other times they come unexpectedly; moreover, since no unique mechanism for nucleation was proven to exist, it is unlikely that single classical precursors (e.g., increasing seismicity, geochemical anomalies, geoelectric potentials) may ever be effective in predicting impending earthquakes. For this reason, understanding the physics driving the evolution of fault systems is a crucial task to fine-tune seismic prediction methods and for the mitigation of seismic risk. In this work, an innovative idea is inspected to establish the proximity to the critical breaking point. It is based on the mechanical response of faults to tidal perturbations, which is observed to change during the "seismic cycle". This technique allows to identify different seismic patterns marking the fingerprints of progressive crustal weakening. Destabilization seems to arise from two different possible mechanisms compatible with the so called *preslip patch*, *cascade* models and with seismic quiescence. The first is featured by a decreasing susceptibility to stress perturbation, anomalous geodetic deformation, and seismic activity, while on the other hand, the second shows seismic quiescence and increasing responsiveness. The novelty of this article consists in highlighting not only the variations in responsiveness of faults to stress while reaching the critical point, but also how seismic occurrence changes over time as a function of instability. Temporal swings of correlation between tides and nucleated seismic energy reveal a complex mechanism for modulation of energy dissipation driven by stress variations, above all in the upper brittle crust. Some case studies taken from recent Greek seismicity are investigated.

Keywords: tidal triggering of earthquakes; seismic cycle; coulomb failure stress; preparatory phase; seismic prediction

Citation: Zaccagnino, D.; Telesca, L.; Doglioni, C. Different fault response to stress during the seismic cycle. *Appl. Sci.* **2021**, *11*, 9596. https://doi.org/10.3390/app11209596

Academic Editor: Igal M. Shohet

Received: 6 September 2021
Accepted: 11 October 2021
Published: 14 October 2021

1. Introduction

Experimental and numerical simulations show that disorder plays a key role in driving stress accumulation in the crust and energy nucleation during earthquakes [1], nevertheless it was not clarified yet how stress variations trigger breaking processes in such heterogeneous media. Indeed, earthquakes can be due to several stress sources, such as magmatic intrusion or overpressured liquids; moreover, faulting is also affected by temperature, confining and pore pressure, and rock brittleness. This is why the comprehension of the response of faulting to additional stress was so actively investigated for 50 years. There are a few exogenous stress sources useful for this purpose: fluid injection is a widespread technique in stimulating production from oil and natural gas wells and improve geothermal energy generation. Although it is usually associated with microseismicity, several events with moderate magnitudes were also related to this practice [2]. This is why it is of paramount importance to improve our knowledge about the conditions under which intermediate magnitude events might occur. Also, it is crucial to note that injection and depletion generally happen at different wells, leading to a complex underground liquid circulation. Thus, it is not so easy to model how fluid injection drives spatial variations in pore pressure. An additional source of complexity is due to the variability of the time interval between the beginning of fluid injections and the onset of seismic activity. Therefore, fluid injection cannot be considered an efficient way of monitoring fault response to stress perturbation, at least over the time interval we are interested in. For this reason, we do not focus on this kind of stress source. A second possibility may be the controlled use of explosive, a tested tool in engineering of rock blasting, drilling, and mining, and moreover it is at the base of field reflection and refraction seismology. Unfortunately, this method is useful only for stress pulses simulations.

On the contrary, lunar and solar tides continuously induce periodic deformations in solid earth. Tidal harmonics are featured by different frequencies, so that their periods range from 10^4 to 10^9 s. The displacement of the tidal bulge can be decomposed into its vertical and horizontal components that depend on latitude, and may amount respectively up to 40 cm and 20 cm in the case of the semidiurnal M_2 tide. Solid Tides depend on depth reaching their highest intensity around 1000 km below the surface [3]. Despite the fact that solid and ocean tidal stress (0.1 kPa–100 kPa) is fairly smaller than the earthquake stress drops (1 MPa–30 MPa, [4]) it was sufficiently proven that tides can trigger earthquakes (e.g., [5–9]) even though global seismicity weakly correlates with the Moon and Sun's distances from Earth and long catalogs ($\geq 10^4$ events, [10]) are needed to detect this effect accurately. For these reasons, in this article we investigate how the response of faults to stress modulations changes during the "seismic cycle" using tidal perturbations. In particular, we focus on recent Greek seismicity.

Since the tidal bulge is misplaced relative to the gravitational Earth-Moon alignment, being about 0.3–2.4 degrees eastward of it [11,12] due to the delay in reaction for the anelastic component of the Earth as a response to the tidal pull, a westerly-directed horizontal drag acts on the lithosphere and the continuously slows down Earth rotation [13]. The crux of the matter is that, besides the symmetric oscillatory motions, the nonlinear response of the low velocity zone (LVZ) breaks the symmetry of the tidal traveling wave on the Earth surface: the small asymmetry produces a net drift motion of any material point interacting with the gravitational wave force, in the direction following the rotation of the Moon. Therefore, detailed properties of the combined tidal oscillation and tidal drift depend on the degree of deformability of the lithosphere (a.k.a. Love and Shida numbers) due to its local temperature and geochemistry. These modulations act through two different mechanisms on the outer layers of the Solid Earth: the brittle and outermost part of the crust is elastically affected by tidal waves, which induce a stress variation in rocks that, depending on the local geodynamics, promotes or, on the contrary, can prevent the achievement of the critical breaking point, so that it plays a statistically significant role in fault activation [14–16], while solid tides actively modulate plate motion at low frequencies over geological periods [17,18] due to a heterogeneous dissipation of the tidal torque at the LVZ level.

The issue of the tidal triggering of seismicity is not trivial, since it is necessary to take into account not only of the effect of Earth tides, but also variations in pore and confinement pressure, the orientation of the DC (Double Couple contribution) in relation with focal mechanism, and CLVD (Compensated Linear Vector Dipole) of earthquakes and local geophysical heterogeneities.

To make matters even more complicated, a further difficulty must be considered: the observed seismicity spans up to 11 orders of magnitude over time (from 10 s typical of microseismicity up to 10^{12} s for earthquakes recurrence time intervals), and 14 concerning to energy (if one considers events ranging from M_w 0 to 9.5) and seven in space (a M_w 9.0 can breaks crust up to 1000 km); this creates an enormous obstacle of both resolution and saturation, catalog incompleteness [19], and unreliability of statistical data.

Finally, to the technical complexity of measuring real stress, we add the estimation of uncertainties of seismic parameters such as magnitude and depth.

Beyond the triggering mechanism, in this work we focus on the possibility of highlighting the growth of critical states in the crust induced by stress accumulation in rocks through the measurement of correlations between some features of the tidal perturbation and seismic activity.

There is a flurry of scientific articles devoted to understand how tides influence seismic activity, but only a few of them (e.g., [20]) so far extensively studied whether tidal perturbation might somehow provide information about the stability of faults close to rupture, with a few exceptions regarding some particular case studies (e.g., [21,22]). The novelty of this article with respect to previous scientific literature consists in highlighting not only the variations in responsiveness of faults to stress while reaching the critical point, i.e., before a large earthquake, but it also shows how seismic response of faults changes over time as a function of their instability.

2. Materials and Methods

The lunisolar tides deform the Earth up to 60 cm twice a day, moreover the weight of the ocean tides gives a periodic load on the Earth's surface strongly dependent on bathymetry [23]. Although the displacements are relatively large, the associated changes in strain at the Earth's surface are tiny, extremely difficult to measure accurately, and even more tricky to model. The main difference between liquid tides and solid tides is in the phase: rocks react quickly to solicitations, while fluid masses need a characteristic time span to move, so they are affected by the tide with a certain phase shift.

We model tidal stress according to the following method.

Considering two massive celestial bodies with a spherical distributed mass density one realizes that the gravitational and centrifugal forces are balanced whenever their volumes are deformed by tides.

$$\vec{F}_{\mathcal{M}}(P) = \vec{F}_G(P) + \vec{F}_C(P) \tag{1}$$

from which the tidal acceleration is immediately obtained

$$|\vec{a}_{\mathcal{M}}(R)| = \frac{GM}{(R-r)^2} - \frac{GM}{R^2} \simeq \frac{2MGr}{R^3} \tag{2}$$

in the last step $R \gg r$ is assumed. The gravitational potential due to celestial body with mass M, at a point P at a distance r from the center of the planet, can be expanded in a series of powers of r/R, where R is the distance from the center of the Earth to the celestial body.

$$W(P) = V(P) + \vec{F}_G(P) \cdot \vec{r}(P) + const \tag{3}$$

$$W(\vec{R}) = -\frac{GM}{|\vec{R}+\vec{r}|} + \frac{GM}{R^3}\vec{R} \cdot \vec{r} - \frac{GM}{R} \tag{4}$$

if $l = |\vec{R}+\vec{r}|$

$$W(\vec{R}) = -\frac{GM}{R}\left(\frac{R}{l} - \frac{\vec{R} \cdot \vec{r}}{R^2} - 1\right) \tag{5}$$

by expanding in series of Legendre polynomials we obtain

$$W(R, \Psi) = \frac{GM}{R} \sum_{n=2}^{\infty} \left(\frac{r}{R}\right)^n P_n(\cos \Psi) \tag{6}$$

Since r/R for the Moon is $\sim 1/60$ while for the Sun is $\sim 1/23{,}000$, the contributions of successive terms to the potential rapidly decrease. For the Moon, $\sim 98\%$ of the total tidal potential, and for the Sun, the higher orders are completely negligible for our purpose. Moreover, from the ratios of masses and their mean distances, it follows that the solar tidal perturbation is ~ 0.459 times the lunar tides. So, we can write that the total tidal potential is given by the sum of lunar and solar perturbations in the following form

$$W(R, \Psi) \simeq \frac{GM}{R} \left(\frac{r}{R}\right)^2 P_2(\cos \Psi) \tag{7}$$

where Ψ is the zenith of the body with respect to P and P_2 is the second degree Legendre polynomial.

$$cos\Psi = \cos \theta \cos \delta + \sin \theta \sin \delta \cos(\phi - \alpha) \tag{8}$$

θ is the colatitude and ϕ is the easterly longitude of P, δ is the codeclination, and α is the right-ascension of the body. We can write the potential so that three different contributions are highlighted

$$W(R, \Psi_1, \Psi_2, \Psi_3) = \frac{3GMr^2}{4R^3} [\Psi_1 + \Psi_2 + \Psi_3] \tag{9}$$

with

$$\Psi_1 = 3\left(\sin^2\left(\frac{\pi}{2} - \theta\right) - \frac{1}{3}\right)\left(\sin^2\left(\frac{\pi}{2} - \delta\right) - \frac{1}{3}\right) \tag{10}$$

$$\Psi_2 = \sin\left(2\left(\frac{\pi}{2} - \theta\right)\right)\sin\left(2\left(\frac{\pi}{2} - \delta\right)\right)\cos(\phi - \alpha) \tag{11}$$

$$\Psi_3 = \cos^2\left(\frac{\pi}{2} - \theta\right)\cos^2\left(\frac{\pi}{2} - \delta\right)\cos(2(\phi - \alpha)) \tag{12}$$

For the sake of simplicity

$$D = \frac{3GMr^2}{4R^3}$$

is called Doodson's parameter.

The vertical displacement is obtained by dividing W for the local value of the gravitational acceleration

$$\delta\mathcal{H}(R, \Psi_1, \Psi_2, \Psi_3) = \frac{W(R, \Psi_1, \Psi_2, \Psi_3)}{g} = \frac{3GMr^2}{4gR^3}\left(\Psi_1 + \Psi_2 + \Psi_3\right) \tag{13}$$

The three terms represent the zonal, tesseral, and sectoral tides respectively. The zonal and sectoral contributes are responsible for tides with a half the Moon's revolution period, while the tesseral one for planet's rotation period tides. A paramount research work in this field was conducted by A. T. Doodson (1890–1968), who identified 378 tidal harmonics that were collected in a celebrated catalog in 1921 [24]. The largest tidal harmonics [25] are shown in Table 1. The elastic deformation of the Earth has two modes: spheroidal and toroidal, but tidal forces excite only the spheroidal modes [26].

Table 1. Largest tidal frequencies.

Symbol	Doodson Number	Period (Days)	Amplitude (m)
M_2	2 5 5 5 5 5	0.518	0.6322
S_2	2 7 3 5 5 5	0.500	0.2941
K_1	1 6 5 5 5 5	0.997	0.3686
M_f	0 7 5 5 5 5	13.661	$-$ 0.0666
M_m	0 6 5 4 5 5	27.555	$-$ 0.0352
SSa	0 5 7 5 5 5	182.622	$-$ 0.0310
h	0 5 6 5 5 4	365.264	$-$ 0.0049
p	0 5 5 6 5 5	3232.605	0.0002
N/2	0 5 5 5 7 5	3399.048	$-$ 0.0003
N	0 5 5 5 6 5	6798.097	0.0279

For the sake of simplicity, we assume that the Earth is spherically symmetric, nonrotating, elastic, and isotropic. Nevertheless, the effect of sphericity of the Earth's layering cannot be neglected if one wishes to calculate surface waves of long wavelength. Spheroidal deformations of the SNREI model as a function of depth, Lamé's coefficients and the gravitational acceleration can be evaluated by using a set of functions y_i with $i = 1, \ldots, 6$ which satisfy a set of six ordinary differential equations [27]

$$\frac{dy_i(r,n)}{dr} = \sum_{j=1}^{6} f(\rho(r), \lambda(r), \mu(r), g(r), n) y_j(r,n),$$

(14)

where n is the n-th mode. For our purpose, it is enough to consider $n = 2$. Displacements can be written in spherical coordinates as follows:

$$\begin{cases} u_r = \dfrac{h_2(r)}{g(r)} W(r,\theta,\phi) \\ u_\theta = \dfrac{l_2(r)}{g(r)} \dfrac{\partial W(r,\theta,\phi)}{\partial \theta} \\ u_\phi = \dfrac{l_2(r)}{g(r)\sin\theta} \dfrac{\partial W(r,\theta,\phi)}{\partial \phi} \end{cases}$$

(15)

The value of $g(r)$ rises from the surface up to a depth of about 700 km to a maximum of 9.99 m/s^2. In the lower mantle, $g(r)$ lingers stable and increases abruptly near the Gutenberg discontinuity, reaching 10.16 m/s^2. Gravity continuously falls in the core with a rate that depends on density till it reaches a zero-value at the center of the Earth.

In this work, tidal perturbations are considered with respect to seismicity and, above all, crustal seismicity (up to $\sim$100 km at depth), so that if we assume $g(r) \simeq g(r = 6371$ km$)$ an error $\sim$0.1% is introduced, which is negligible. The strain components in spherical coordinates are obtained from (15) by derivation

$$\begin{cases} \varepsilon_{rr} = \dfrac{\partial u_r}{\partial r} \\ \varepsilon_{\theta\theta} = \dfrac{1}{r}\left(\dfrac{\partial u_\theta}{\partial \theta} + u_r \right) \\ \varepsilon_{\phi\phi} = \dfrac{1}{r\sin\theta}\left(\dfrac{\partial u_\phi}{\partial \phi} + u_r\sin\theta + u_\theta\cos\theta \right) \\ \varepsilon_{\theta\phi} = \dfrac{1}{r}\left(\dfrac{1}{\sin\theta}\dfrac{\partial u_\theta}{\partial \phi} + \dfrac{\partial u_\phi}{\partial \theta} - u_\phi\cot\theta \right) \end{cases}$$

(16)

For computational simplicity, the radial component of strain can also be evaluated by using

$$\varepsilon_{rr} = -\frac{\nu}{1-\nu}\left(\varepsilon_{\theta\theta} + \varepsilon_{\phi\phi} \right)$$

(17)

where ν is the Poisson's coefficient that can be computed at depth with

$$\nu(r) = \frac{\lambda(r)}{2(\lambda(r) + \mu(r))} \tag{18}$$

in turn, the Lame's coefficients can be obtained starting from the speed of the seismic P and S waves as a function of depth (PREM, [28]).

$$\begin{cases} \lambda(r) = \rho(r)\left(v_P^2(r) - v_S^2(r)\right) \\ \mu(r) = \rho(r)v_S^2(r) \end{cases} \tag{19}$$

So, the needed components of stress in spherical coordinates are

$$\begin{cases} \sigma_{\theta\theta}(r) = \lambda(r)\left(\varepsilon_{rr} + \varepsilon_{\phi\phi} + \varepsilon_{\theta\theta}\right) + 2\mu(r)\varepsilon_{\theta\theta} \\ \sigma_{\phi\phi}(r) = \lambda(r)\left(\varepsilon_{rr} + \varepsilon_{\phi\phi} + \varepsilon_{\theta\theta}\right) + 2\mu(r)\varepsilon_{\phi\phi} \\ \sigma_{\theta\phi}(r) = \mu(r)\varepsilon_{\theta\phi}. \end{cases} \tag{20}$$

At last, it is necessary to take into account the spatial orientation of faults, which provides information about the tectonic stress tensor. Given the strike α of the seismological source, the tangential stress is

$$\sigma_\alpha^{(\pm)} = \sigma_{\theta\theta}(r)\cos^2\alpha + \sigma_{\phi\phi}(r)\sin^2\alpha \pm 2\sigma_{\theta\phi}(r)\sin\alpha\cos\alpha. \tag{21}$$

Then, the quantities of geophysical interest are the following

$$\begin{cases} \sigma_s = \sigma_\alpha^{(+)}\sin\delta\cos\delta \\ \sigma_n = \sigma_\alpha^{(+)}\sin^2\delta \\ \sigma_c = \frac{1}{3}\left(\sigma_\alpha^{(+)} + \sigma_\alpha^{(-)}\right) \end{cases} \tag{22}$$

where δ is the dip angle of the fault. α and δ are inferred by comparing the focal mechanisms of local seismicity with the maps of actual faults. Since focal mechanisms are calculated only for earthquakes with significant magnitude, usually larger than 3.5–4.0, routinely recorded small events are assumed to occur on fault planes whose angles of strike and dip are given by the averages of the available ones.

σ_s is the so-called shear stress, which is positive in extensional tectonics, σ_n is the normal stress acting orthogonally to the fault plane, and σ_c is the confining stress due to the weight of the overlying rocks and fluids [29]. σ_{rr}, $\sigma_{r\theta}$, and $\sigma_{r\phi}$ are not considered since they are negligible up to 300 km deep [3] and $\sim$95% of the seismic energy is nucleated within the depth 0–50 km. For these reasons, only the horizontal shear stresses $\sigma_{\theta\theta}$ and $\sigma_{\phi\phi}$ can effectively play a role in triggering earthquakes.

A still open problem concerns the functions y_i: for the calculations above, only y_1 and y_3 are needed since

$$\begin{cases} y_1(r,2) = \dfrac{h_2(r)}{g(r)} \\ y_3(r,2) = \dfrac{l_2(r)}{g(r)} \end{cases} \tag{23}$$

they can be obtained by integrating with the fourth order Runge–Kutta method, a system of six coupled ordinary differential equations starting from a set of suitable boundary conditions. In turn, these can be calculated by considering the solutions in the case of an isotropic and homogeneous sphere on the Gutenberg discontinuity. The result is a combination of spherical Bessel function of the first kind that can be computed with a power series expansion.

As regards ocean tides, a foreword is necessary.

Despite ocean loading being able to induce stress up to 100 kPa, which is much larger than the stress due to solid tides (0.1–3 kPa), it is locally generated and usually focused over small surfaces ($\lesssim 10^4$ km^2), with some exceptions such as off the coast of New Zealand, the Madagascar Channel, the Java–Timor Sea and offshore Alaska. In practice, the main contribution of oceanic tides derives, unlike solid tides, from vertical stress [30]

$$\sigma_{zz} = -\rho g h \tag{24}$$

where h is the amplitude of the tide and $\rho \simeq 1030$ kg/m^3; the radial stress spread horizontally through the Poisson's coefficient. So, working in local Cartesian coordinates, if I assume that the vertical stress acts symmetrically $\sigma_{xx} = \sigma_{yy}$ [31]

$$\sigma_{xx} = \frac{\nu}{1-\nu}\sigma_{zz} \tag{25}$$

therefore, comparing with Equation (22) we get

$$\begin{cases} \sigma_s \approx 0 \\ \sigma_n \approx \sigma_{xx}\cos^2\delta + \sigma_{yy}\sin^2\delta \\ \sigma_c = \dfrac{1}{3}(\sigma_{xx} + \sigma_{yy} + \sigma_{zz}) \end{cases} \tag{26}$$

the predicted tidal height H can be provided by the NAO.99b software [32].

At last, it is convenient to introduce the Coulomb Failure Stress (CFS) [33]

$$\text{CFS} = |\sigma_{ts}| - \mu(\sigma_{tn} - p) - S_0 \tag{27}$$

where σ_{ts} and σ_{ns} are, respectively, the shear and normal stress, p is the pore pressure and S_0 stands for the coesion of rocks. It is often assumed that changes in p are proportional to the normal stress change across the fault plane, so that rescaling μ

$$\text{CFS} = \sigma_{ts} - \mu\sigma_{tn} - S_0 \tag{28}$$

with $\mu \sim 0.4$–0.8.

Since the original state of stress is unknown, the ΔCFS [34] is usually studied

$$\Delta\text{CFS} = \sigma_s + \mu\sigma_n \tag{29}$$

where σ_s is the change in shear stress on the fault plane induced by the tidal perturbation in the slip direction and σ_n is the tidal normal stress. Positive ΔCFS is associated with encouraged seismicity, while negative values produce stress shadow effects which inhibit slip [35].

Since tidal stress upon the fault is known in principle, we can perform a correlation analysis according to the following steps:

- Identification of regions of geophysical interest based on recent seismic events. The ranges of latitude, longitude, and depth are selected for each area;
- The completeness magnitude is estimated for each catalog;
- Declustering is carried out following the method proposed by Uhrhammer, to remove seismic sequences generated by events with high magnitude. Retrospectively, the role of declustering is studied. Banking on our results discussed in the next section, our analysis should be performed without declustering;
- M_L and M_b are converted into moment magnitudes M_w;
- Normal, shear, confinement and ΔCFS stresses are calculated for each earthquake occurred within the selected region following the procedure above;
- Uncertainties on the stress values are estimated by propagation of the errors in the measure of spatial parameters of faults and focal mechanisms and hypocentral pa-

rameters. The dominant contribution comes from the strike and dip angle errors, so that

$$
\begin{cases}
\epsilon_{shear} \simeq \sqrt{\epsilon_+^2 \sin^2 \delta \cos^2 \delta + \epsilon_\delta^2 \sigma_+^2 \cos^2 \delta} \\
\epsilon_{normal} \simeq \sqrt{\epsilon_+^2 \sin^4 \delta + \epsilon_\delta^2 \sin^2 2\delta} \\
\epsilon_{confinement} \simeq \frac{1}{3}\sqrt{\epsilon_+^2 + \epsilon_-^2} \approx 0.47\epsilon_+
\end{cases}
\tag{30}
$$

where ϵ replaces the usual symbol for standard deviation to avoid misunderstanding with stress components σ_s, σ_c, σ_n and $\sigma_\alpha^{(\pm)}$.
ϵ_+ and ϵ_- are the uncertainties of the positive and negative tangential stresses given by

$$
\epsilon_+ = \epsilon_- \simeq \epsilon_\alpha \sqrt{(2\sigma_{\theta\theta} \cos\theta \sin\theta)^2 + (2\sigma_{\phi\phi} \cos\theta \sin\theta)^2 + (2\sigma_{\theta\phi} \cos 2\theta)^2}
\tag{31}
$$

- The correlation between the magnitude of the seismic events and the intensities of the tidal stress components acting on the fault is calculated over fixed time intervals Δt according to the following formula:

$$
\rho_{t_n}^{(j)} = \frac{\sum_{i=1}^{N_{t_n}} \left(M_{wi} - \overline{M_w}\right)\left(\sigma_i^{(j)} - \overline{\sigma^{(j)}}\right)}{\sqrt{\sum_{i=1}^{N_{t_n}} \left(M_{wi} - \overline{M_w}\right)^2 \sum_{k=1}^{N_{t_n}} \left(\sigma_k^{(j)} - \overline{\sigma^{(j)}}\right)^2}}
\tag{32}
$$

where $j = s, n, c, \Delta CFS$ meaning respectively the shear, normal, confinement and ΔCFS components of stress; N_{t_n} is the number of failures occurred during the n-th time step.
ΔCFS is calculated for each event occurred within the selected region whose magnitude is above the completeness magnitude. Since we are interested in understanding how sensitivity of faults to additional stress modulations changes during the seismic cycle, we calculate the correlation between ΔCFS and the nucleated seismic energy for several time intervals (about 20 in Figures 1–3). To do this, the number of events used for the calculation of the correlation must be large enough to suppress stochastic fluctuations (the number of earthquakes for each point in Figures 1–3 is >200). On the other hand, short time intervals are not suitable for our goal because tides with not negligible amplitudes have semiannual and yearly frequencies. Therefore, $\Delta t < 1$ yr can affect the correlation value. Moreover, we are looking for slow processes of progressive destabilization of crustal volumes, then averaging does not cause information loss, but only noise attenuation.
- The uncertainty on the correlation index is obtained by propagation of tidal stress errors and magnitudes.

3. Results

What happens before large stress drop? In the nucleation dominated regime [10], the mechanism responsible for fracture triggering is analogous to static fatigue and delayed failure. The correlation between earthquake occurrence and tidal phase vanishes and failure is ultimately controlled by stress maxima. This can appear to be contradictory because we underlined that tidal stress modulations are $\sim 10^4$ smaller than the seismic stress drop. Mechanical triggering must play a key role in statistical seismology, so that a tiny initial perturbation always have a not vanishing probability to become an earthquake. In this view, the problem of triggered instability reduces to a trivial threshold phenomenon in which perturbations are just "the straws breaking the camel's back".

Analogously to mechanical engineering studies delving into the periodic supervision of facilities to detect signals of progressive weakening or corrosion, it is possible that seismicity could show significant variations in the correlation between seismic activity and stress modulations before a major event. Below we analyze three among the most important seismic sequences recorded in Greece in the last 20 years.

Greece is a country prone to elevated seismic risk with both continental and insular territories prone to large earthquakes. Since in the Mediterranean Sea the amplitude of liquid tides does not reach relevant values (<50 cm), and they are even smaller in the investigated region, i.e., Greek Ionian Sea, where M_2 tide is about 5 cm high [36], sea tides can be neglected for our purpose.

In our analysis, we focus on three different regions: Northern Thessaly Region, Northern Ionian Greek Islands, and Southern Ioanian Greek Islands.

Northern Thessaly Region was recently hit by the Larissa seismic sequence (mainshock 3 March 2021, M_w 6.3, 11.5 km depth, USGS), causing widespread damage and one casualty. The still ongoing seismicity occurred along normal faulting. The analysis of correlation between the ΔCFS and the nucleated seismic energy is performed between 1990 and 1 May 2021 in an area within latitude 39.5–40.2° N and longitude 21.7–22.4° E considering only earthquakes with $M_L > 2.0$. After an initial period featured by elevated correlation values corresponding to a seismic swarm ($M_w \leq 5.0$, $\rho \sim 0.45$) which occurred between 1993 and 1995, a decennial decrease of correlation followed. The trend switched in 2005–2007 and continued till the correlation turns positive. The progressive increase stopped in 2010 when diffuse swarms were recorded.

Then, tidal correlation becomes negative again till 2020, when a peak is rapidly reached at the beginning of 2021 ($\rho \sim 0.29$), when Thessaly was shaken by a M_w 6.3 earthquake. Our results are summarized in Figure 1.

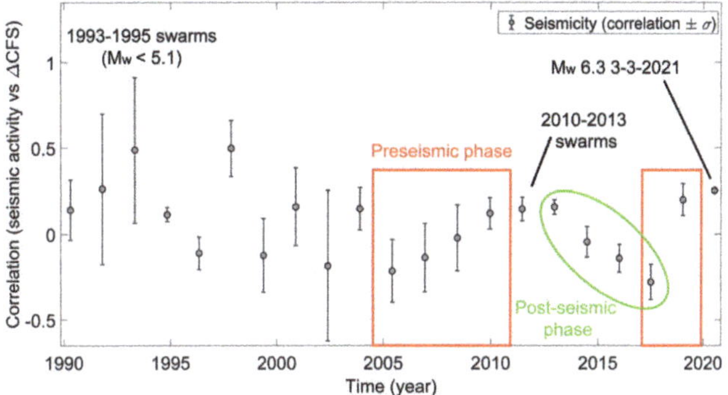

Figure 1. Correlation between ΔCFS and seismicity in Thessaly Greece, between 1990 and 2021, $M_L > 2.0$, NOAIG Catalog.

Northern Ionian Greek Islands are often involved by seismic sequences because they rise along the regional plate boundary between the Africa and Eurasia plates, which converge at a rate of about 9 mm/yr towards the north-north west.

Nubian lithosphere subducts beneath the Aegean Sea along the Hellenic Arc.

Focal mechanisms clearly denote compressive earthquakes. The largest seismic events during since 1990 occurred on 17 November 2015, M_w 6.5 [37], and it is usually known as the Lefkada earthquake because it resulted in several fatalities and dozens injuries on the Greek Island of Lefkada.

This major seismic crisis was forerun by a two-years-long seismic instability, which also included large-magnitude events such as the Cephalonia earthquake (26 January 2014, M_w 6.1).

Correlation analysis shows an about five years lasting increase of ρ, which reached its maximum in 2013 ($\rho \sim 0.13$), then a sharp fall is observed so that tidal correlation becomes compatible with zero. Our results are in Figure 2.

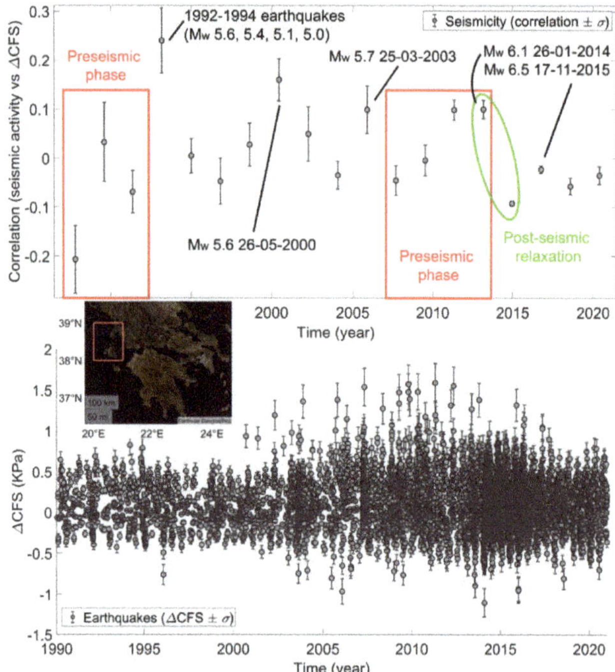

Figure 2. (**Top**) correlation between ΔCFS and seismicity close to the Greek Ionian coasts (38.0–39.0° N, 20.0–21.0° E, $M_L > 2.0$, NOAIG Catalog.) between 1990 and 2021. Main event was located in southwestern region of Lefkada Island between the villages Athani and Agios Petros. The 2014–2016 seismic sequence involved faults within different tectonic settings. (**Bottom**) ΔCFS as a function of time. Before the 2014–2016 seismic sequence, significant changes in the Coulomb stress distribution was observed, partially due to contribution of deeper (>20 km) earthquakes.

The tectonic setting of the Southern Ioanian Greek Islands (37.0–38.0° N, 20.0–21.0° E) is similar to the previous one.

Great part of seismicity is still featured by a compressive focal mechanism even if a fraction of crustal events with significant oblique component are recordered.

In this territory, two main quakes happened in the last thirty years.

The largest hit Lithakia on 25 October 2018 with M_w 6.8, depth 15 km, it also produced a small tsunami with about ∼20 cm high anomalous waves.

The other was a M_w 6.6 earthquake occurred in the same area on 18 November 1997.

The Lithakia earthquake was preceded by an about 7–9 years long period of progressively increase of the tidal correlation between ΔCFS and seismic energy. The peak of the correlation was measured between 2018 and 2019 with $\rho \sim 0.11$, then a fast decrease is observed.

At the same time, an anomaly in the localization of seismicity is noticed, which is reported in the lower part of Figure 3. Two seismic shadow zones are located at a depth of 0–10 km (2012–2017) and below 19 km (2015–2018).

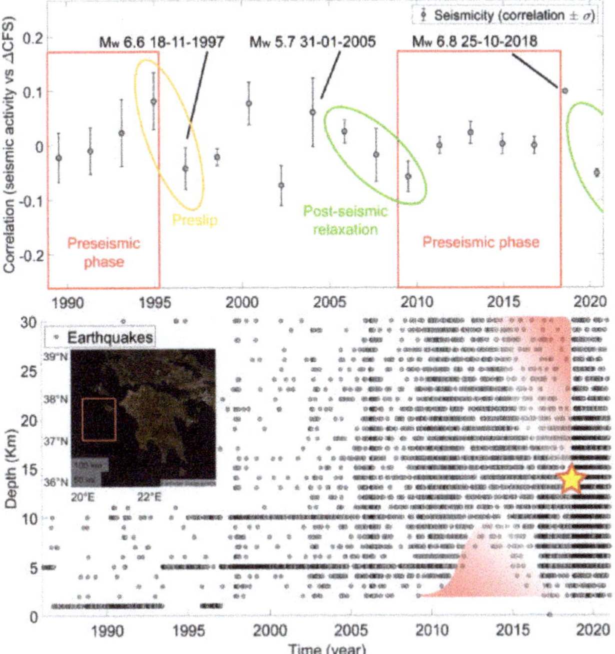

Figure 3. Correlation between ΔCFS and seismicity close to the Greek Ionian coasts (37.0–38.0° N, 20.0–21.0° E) between 1985 and 2021. A nine-years-long preseismic phase is highlighted both by an increasing value of correlation in the upper part of the picture, and seismic quiescence, represented by two red shadow zones, at a depth of 0–10 km (2012–2017) and >20 km (2015–2018). The yellow star represents the M_w 6.8, 25 October 2018 Lithakia mainshock.

The seismic quiescence [38] in the forementioned layers is statistically significant especially in the upper one, where the reduction of the seismic rate reached 60% with respect to the preceding 2007–2012 period.

Observations suggest that ρ tends to increase during the preparatory phase of significant crustal earthquakes in Greece. The values of correlation are higher in extensional tectonic settings rather than compressive ones of about 100–200%, normal fault quakes are found to correlate stronger with tidal stress modulations also in other regions. The energy conditions are instead stricter for events occurring in a compressive tectonic setting, so it is reasonable to expect that they correlate very little with the intensity of tidal perturbations, especially in the case of deep hypocenters. Therefore, even though correlations are weak, their modulations can provide precious information about the condition of instability of local crustal volumes, especially if jointly analyzed with other seismological and geodetic recordings. We also analyze the spatial density of the Coulomb stress variation induced by the action of tidal perturbations. For each seismic event, ΔCFS value is calculated, then a map is created that shows the fraction of ΔCFS generated in each location of the selected region. Therefore, the areas with elevated ΔCFS density are those in which seismicity has statistically occurred at more elevated Coulomb stress values or characterized by higher seismicity rate with respect to the surrounding areas.

Since the seismic rate is ultimately controlled by the maximum nucleated magnitude, for each case study, we also plot the areas hit by earthquakes with large magnitude, as written in the caption according to the relative intensity of the regional seismicity. The green contours are drawn according to the finite fault maps of the USGS catalog. Whenever earthquake sequences occur at significantly high ΔCFS values, the brightest spots (according to the vertical colorbar on the left in Figure 4) are located within the green profiles, which suggest self-triggering. On the contrary, shiny stains just outside the main shock areas point out zones featured by elevated stress transfer.

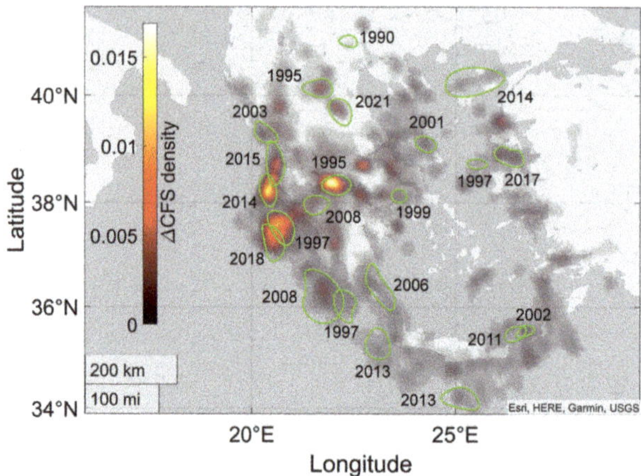

Figure 4. ΔCFS density map for seismicity in Greece, NOAIG Catalog, 1990–2021, $M_L \geq 2.0$. Highest density areas are located on the Greek Ionian Islands along Hellenic Trench and along normal fault system of Gulf of Corinth.

The ΔCFS density map for Greece shows diffuse signal sometimes due to strong motion recordings (e.g., Gulf of Corinth seismic sequence, 1995, and Ionian Arc seismicity in 1997); nevertheless, no correlation is found between the intensity of the signal and large magnitudes, as proven in the cases of the Methoni M_w 6.9 earthquake and the Aegean Sea M_w 6.9 event.

The brightest patches are located where the 2014–2015 Ionian and the 2018 Lithakia seismic sequences occurred. This means that seismic activity in the Aegean region is usually self-triggered at large spatial scales. Compare with Figure 4 and Figure 5.

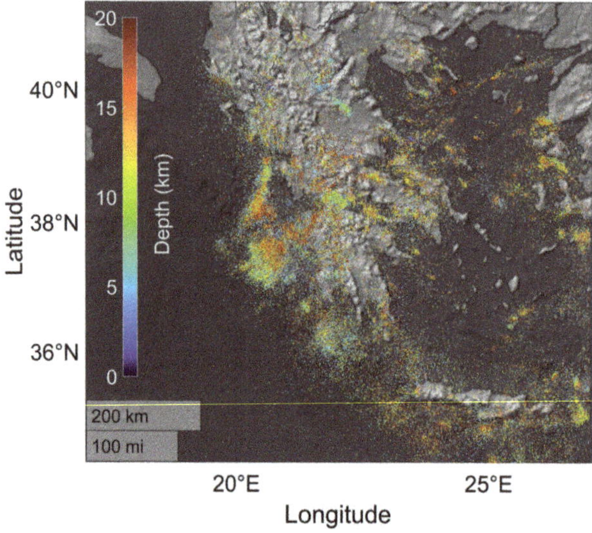

Figure 5. Map of seismicity in Greece between 1990 and 1/8/2021, M_L > 2.0, NOAIG Catalogue.

4. Discussion

Tidal triggering of earthquakes is still a debated theme in Solid Earth Geophysics. In particular, there are three sources of discussion among geophysicists:

- From a statistical viewpoint, earthquake catalogs are often insufficient to detect significant modulations of seismic activity over time.
  Tides are tiny perturbations of the gravitational field (0.1–100 kPa) with respect to typical earthquake stress drops (1–50 MPa), but nonetheless they are able to generate significant stress variation rates ($\sim$100 mbar/day compared with 1–10 mbar/day due to tectonic stress, [39]). However, the actual impact on the stability of rock volumes largely depends on the tectonic setting, the spatial orientation of the fault, the depth, and the hypocentral latitude; finally, also the magnitude of the impending event modifies the response of the system to the tidal perturbation. Therefore, a wide range of results were found in several geographical regions.
- It is sometimes difficult to distinguish between the effects of Solid Earth tides from those of ocean tides. Even though the stress amplitude of liquid tides (up to 100 kPa) often exceeds that of the Solid Earth tide (usually $\sim$0.1–1 kPa), the mechanism of action is quite different. In the first case, tides are concentrated on limited surfaces, at most of the order 10^5 km^2, and act mainly through the σ_{rr} vertical component, which is transmitted to the horizontal components (almost symmetrically) thanks to the elastic properties of the lithosphere. However, the incremental stress decreases exponentially as depth increases, therefore it can strongly modulate shallow oceanic small magnitude seismicity, for example, at ocean ridges or submarine volcanoes, but it is unlikely that intermediate and deep earthquakes might be triggered by liquid tides.
  Solid tides, on the other hand, deform the outer layers of the planet mainly in the horizontal components, acting on very large surfaces. For this reason, solid tides have a dominant role in triggering earthquakes except for the just cited peculiar cases. For these reasons, liquid tides are neglected in this work.
- Seismic response to tidal loading strongly depends on the duration of earthquake nucleation.

Beyond the aforementioned issues, well-established scientific evidence exists about tidal synchronization in seismic catalogs, as already discussed in the introduction. Both global and regional seismic series show semiannual, annual, biennial, with approximately 9-, 19-, 37-, and 56-years-long periods activity modulations. While the first three frequencies are generally associated with seasonal patterns, the others have no explanation other than lunisolar tidal loading. The FFT of European magnitudes (SHEEC 2020 catalogue) in Figure 6 clearly attests this phenomenon. Since $M_c \approx 6$ for the SHEEC catalogue, the nonuniform FFT is applied to include at least $M_{wdef} > 5.0$. Instrumental recordings are rarely available before the 1950s also for violent earthquakes, and therefore, macroseismic intensity data are widely used combined with epicentral macroseismic intensities and other parametric data sources. Therefore, parametric catalogs must be used with caution and results must be interpreted according to their reliability. Even if an accurate analysis cannot be performed for the aforementioned reasons, the power spectrum of the European seismicity shows typical tidal frequencies such as 8–10- or 18–19-years-long periodicities and some multiples. Moreover, local seismic rates are noticed to be directly correlated with the phase of tidal shear stress or the Coulomb failure stress change in submarine volcanic seismicity [31] and seismic tremors [6]. Finally, it is reasonable to expect that the triggering power of tides is affected by the following variables:

- The critical stress needed for crack propagation depends on the tectonic setting. Then normal-fault quakes and oblique flat or low-angle thrust earthquakes are the most sensible to tidal stress changes [40];
- Lithostatic loading increases with a vertical gradient equal to $\sim$27 MPa/km, so that confinement stress requires higher and higher energy activation for fracturing (we

assume to investigate only crustal volumes above the BDT). If pore pressure is neglected, as a first approximation, earthquakes become less and less susceptible both to ocean and to solid tidal loading with increasing depth;

- Tesseral, sectoral, and zonal components of solid tides reach different amplitudes depending on the latitude, and therefore the intensity of the phenomenon is more or less evident depending on the location.

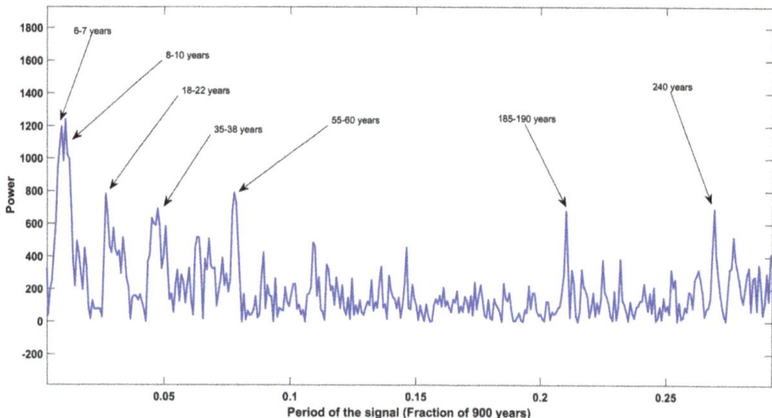

Figure 6. Power spectrum of European seismicity (M > 5.0, 1106–2006, SHEEC catalogue). Tidal periodicities are detected in the recurrence times of intense seismicity in Europe. nuFFT is used for the calculation to take into account progressive decrease of completeness magnitude.

To perform a reliable statistical analysis

$$N \approx \left(\frac{k\sigma_n}{\Delta\sigma_s} \right)^2 \approx 10^3 \tag{33}$$

events are required if we assume $\sigma_n \sim 1$–10 MPa, $\Delta\sigma_s \sim 0.1$–1 kPa and $k \sim 10^{-3}$ (compare with [10], p. 12). This means that only high-quality and extended seismological networks can provide an adequate amount of information for our research. Microseismicity strongly correlate with the phase of tidal loading (e.g., [41]), but we neglected this phenomenon in the present work to focus on tectonic earthquakes. Several studies proved a strong sensitivity of seismicity to stress changes of both endogenous and exogenous origins (e.g., [42]). For this reason, Coulomb failure stress was applied to correlate its variations with changes in aftershocks productivity. A difference between static and dynamic Coulomb stress is conventionally done: when loading is slow, so that its increasing/decreasing rate is negligible with respect to the compared time interval, then the static Coulomb stress is at work, on the contrary, if loading occurs suddenly (i.e., fluid injection, coseismic slip), then the dynamic Coulomb failure stress plays a relevant role. In general, the strain produced by earthquakes induces dynamic Coulomb stress swings that, at long distances, can be even an order of magnitude larger than the static stress changes. There is a nonlinear dependence of the time to instability on stress variations [14]; this not only means that seismic rate is a direct effect of loading, but also implies that small additional stress can result in highly unpredictable states of crustal instability. From a mathematical viewpoint, these properties can be summarized in the seismicity rate $R(t)$ equation, which, in the simplest form, reads [43]

$$R(t) = \frac{R_0}{1 + e^{-\frac{t}{t_A}} \left(e^{-\frac{\Delta CFS}{A\sigma_n}} - 1 \right)} \tag{34}$$

where R_0 is initial seismic rate, $\mathcal{A}$ is a constitutive parameter, and t_A is the duration of the loading.

In brief, observations suggest that seismicity rate can be influenced by both static and dynamic perturbations. If static stress changes act on crustal stability modulating earthquake occurrence, then seismicity rates might be influenced by the Solid Earth tides, caused by the pull of both the Sun and Moon, even though rather weak with respect to tectonic stress. This is the reason why a research looking for tidal static stress loading signatures along the seismic cycle is meaningful and the results showed in this work can be reliable. The case studies we consider suggest that clustered shallow seismicity tends to occur in correspondence with positive values of the correlation ρ between nucleated seismic energy and ΔCFS. The correlation values show progressive growth before major seismic sequences, while they fall while seismicity is ongoing. Preslip, in agreement with [44], and aftershock activity are also both associated with the lowering of correlation values. On the contrary, ρ seems to increase during quiescent periods, which is compatible with [45]. We think that locked faults become more and more sensitive to stress perturbation as they reach the breaking point, which can provide a simple explanation to the observed trends of ρ.

In summary, we develop a method to highlight the different phases of the seismic cycle in fault systems by studying their response to a well-known stress perturbation, i.e., tidal stress. Even though seismic prediction was considered impossible [46], no theoretical reason prevents it. Since the physics of fracture at seismological spatio-temporal scale is still poorly understood and no unique mechanism for nucleation was proven to exist, then seismic precursors cannot be effective in predicting impending earthquakes. By the same token, it is unlikely that the probability of occurrence of single earthquakes may ever be reliably assessed. However, it was proven, also in this article, that fault systems change their mechanical response during the different phases of seismic activity, which is certainly not sufficient for forecasting, but it can be used to understand whether fault systems are evolving towards instability. Our analysis shows that a preseismic phase is observed before large and intermediate ($M_w \gtrsim 5$) shallow (depth ≤ 50 km) earthquakes. Therefore, our advances achieved in this research are significant, with a potential impact on seismic hazards. In addition, they provide new insights for the comprehension of the relationship between stress perturbations, earthquake nucleation, and seismic sequences, which are still to be fully investigated.

Author Contributions: Conceptualization, D.Z. and C.D.; methodology, D.Z. and L.T.; data analysis, D.Z.; writing—original draft preparation, D.Z.; writing—review and editing, C.D. and L.T.; supervision, C.D. and L.T. All authors have read and agreed to the published version of the manuscript.

Funding: This research was funded by ESA through the TILDE Project.

Data Availability Statement: We used free data available at several seismological catalogs, as written inside the text.

Acknowledgments: Discussions with G. Nico, F. G. Panza, F. Ricci, and F. Vespe were precious.

Conflicts of Interest: The authors declare no conflict of interest. This work was developed as part of the TILDE project, ESA. The funders had no role in the design of the study; in the collection, analyses, or interpretation of data; in the writing of the manuscript, or in the decision to publish the results.

References

1. de Geus, T.W.; Popović, M.; Ji, W.; Rosso, A.; Wyart, M. How collective asperity detachments nucleate slip at frictional interfaces. *Proc. Natl. Acad. Sci. USA* **2019**, *116*, 23977–23983. [CrossRef]
2. Ellsworth, W.L. Injection-induced earthquakes. *Science* **2013**, *341*, 1225942. [CrossRef]
3. Varga, P.; Grafarend, E. Influence of tidal forces on the triggering of seismic events. In *Geodynamics and Earth Tides Observations from Global to Micro Scale*; Birkhäuser: Cham, Switzerland, 2019; pp. 55–63.
4. Tanaka, S.; Ohtake, M.; Sato, H. Evidence for tidal triggering of earthquakes as revealed from statistical analysis of global data. *J. Geophys. Res.* **2002**, *107*, 2211. [CrossRef]
5. Heaton, T.H. Tidal triggering of earthquakes. *Geophys. J. Int.* **1975**, *43*, 307–326. [CrossRef]

6. Rubinstein, J.L.; La Rocca, M.; Vidale, J.E.; Creager, K.C.; Wech, A.G. Tidal modulation of nonvolcanic tremor. *Science* **2008**, *319*, 186–189. [CrossRef]

7. Métivier, L.; de Viron, O.; Conrad, C.P.; Renault, S.; Diament, M.; Patau, G. Evidence of earthquake triggering by the solid earth tides. *Earth Planet. Sci. Lett.* **2009**, *278*, 370–375. [CrossRef]

8. Ide, S.; Yabe, S.; Tanaka, Y. Earthquake potential revealed by tidal influence on earthquake size–frequency statistics. *Nat. Geosci.* **2016**, *9*, 834–837. [CrossRef]

9. Kossobokov, V.G.; Panza, G.F. A myth of preferred days of strong earthquakes? *Seism. Res. Lett.* **2020**, *91*, 948–955. [CrossRef]

10. Beeler, N.M.; Lockner, D.A. Why earthquakes correlate weakly with the solid Earth tides: Effects of periodic stress on the rate and probability of earthquake occurrence. *J. Geophys. Res.* **2003**, *108*, 2391. [CrossRef]

11. Munk, W. Once again: Once again tidal friction. *Prog. Oceanogr.* **1997**, *40*, 7–35. [CrossRef]

12. Smith, S.W.; Jungels, P. Phase delay of the solid earth tide. *Phys. Earth Planet. Inter.* **1970**, *2*, 233–238. [CrossRef]

13. Doglioni, C. The global tectonic pattern. *J. Geodyn.* **1990**, *12*, 21–38. [CrossRef]

14. Dieterich, J.H. Nucleation and triggering of earthquake slip: Effect of periodic stresses. *Tectonophysics* **1987**, *144*, 127–139. [CrossRef]

15. Cochran, E.S.; Vidale, J.E.; Tanaka, S. Earth tides can trigger shallow thrust fault earthquakes. *Science* **2004**, *306*, 1164–1166. [CrossRef]

16. Bendick, R.; Mencin, D. Evidence for synchronization in the global earthquake catalog. *Geophys. Res. Lett.* **2020**, *47*, e2020GL087129. [CrossRef]

17. Ide, S.; Tanaka, Y. Controls on plate motion by oscillating tidal stress: Evidence from deep tremors in western Japan. *Geophys. Res. Lett.* **2014**, *41*, 3842–3850. [CrossRef]

18. Zaccagnino, D.; Vespe, F.; Doglioni, C. Tidal modulation of plate motions. *Earth-Sci. Rev.* **2020**, *205*, 103179. [CrossRef]

19. Wiemer, S.; Wyss, M. Minimum magnitude of completeness in earthquake catalogs: Examples from Alaska, the western United States, and Japan. *Bull. Seismol. Soc. Am.* **2000**, *90*, 859–869. [CrossRef]

20. Trotta, J.; Tullis, T. An Independent Assessment of the Load/Unload Response Ratio (LURR) Proposed Method of Earthquake Prediction. *Pure Appl. Geophys.* **2006**, *163*, 2375–2387. [CrossRef]

21. Tanaka, S. Tidal triggering of earthquakes prior to the 2011 Tohoku-Oki earthquake (Mw 9.1). *Geophys. Res. Lett.* **2012**, *39*. [CrossRef]

22. Su, B.; Li, H.; Ma, W.; Zhao, J.; Yao, Q.; Cui, J.; Yue, C.; Kang, C. The Outgoing Longwave Radiation Analysis of Medium and Strong Earthquakes. *IEEE J. Sel. Top. Appl. Earth Obs. Remote Sens.* **2021**, *14*, 6962–6973. [CrossRef]

23. Baker, T.F. Tidal deformations of the Earth. *Sci. Prog.* **1984**, *69*, 197–233.

24. Doodson, A.T. The harmonic development of the tide-generating potential. *Proc. R. Soc. Lond.* **1921**, *100*, 305–329.

25. Agnew, D.C. Earth Tides. In *Treatise on Geophysics, Volume 3: Geodesy*; Elsevier: Amsterdam, The Netherlands, 2010; pp. 163–195.

26. Tsuruoka, H.; Ohtake, M.; Sato, H. Statistical test of the tidal triggering of earthquakes: Contribution of the ocean tide loading effect. *Geophys. J. Int.* **1995**, *122*, 183–194. [CrossRef]

27. Takeuchi, H.; Saito, M. Seismic surface waves. *Method Comp. Phys.* **1972**, *11*, 217–295.

28. Dziewonski, A.M.; Anderson, D.L. Preliminary reference Earth model. *Phys. Earth Planet. Inter.* **1981**, *25*, 297–356. [CrossRef]

29. Lambert, A.; Kao, H.; Rogers, G.; Courtier, N. Correlation of tremor activity with tidal stress in the northern Cascadia subduction zone. *J. Geophys. Res.* **2009**, *114*. [CrossRef]

30. Tan, Y.J.; Waldhauser, F.; Tolstoy, M.; Wilcock, W.S. Axial Seamount: Periodic tidal loading reveals stress dependence of the earthquake size distribution (b value). *Earth Planet. Sci. Lett.* **2019**, *512*, 39–45. [CrossRef]

31. Scholz, C.H.; Tan, Y.J.; Albino, F. The mechanism of tidal triggering of earthquakes at mid-ocean ridges. *Nat. Commun.* **2019**, *10*, 1–7. [CrossRef]

32. Matsumoto, K.; Takanezawa, T.; Ooe, M. Ocean tide models developed by assimilating TOPEX/POSEIDON altimeter data into hydrodynamical model: A global model and a regional model around Japan. *J. Oceanogr.* **2000**, *56*, 567–581. [CrossRef]

33. Oppenheimer, D.H.; Reasenberg, P.A.; Simpson, R.W. Fault plane solutions for the 1984 Morgan Hill, California, earthquake sequence: Evidence for the state of stress on the Calaveras fault. *J. Geophys. Res.* **1988**, *93*, 9007–9026. [CrossRef]

34. Harris, R.A. Earthquake stress triggers, stress shadows, and seismic hazard. *Curr. Sci.* **2000**, *79*, 1215–1225.

35. King, G.C.; Stein, R.S.; Lin, J. Static stress changes and the triggering of earthquakes. *Bull. Seismol. Soc. Am.* **1994**, *84*, 935–953.

36. Arabelos, D.N.; Papazachariou, D.Z.; Contadakis, M.E.; Spatalas, S.D. A new tide model for the Mediterranean Sea based on altimetry and tide gauge assimilation. *Ocean Sci.* **2011**, *7*, 429–444. [CrossRef]

37. Chousianitis, K.; Konca, A.O.; Tselentis, G.A.; Papadopoulos; G.A.; Gianniou, M. Slip model of the 17 November 2015 Mw= 6.5 Lefkada earthquake from the joint inversion of geodetic and seismic data. *Geophys. Res. Lett.* **2016**, *43*, 7973–7981. [CrossRef]

38. Reasenberg, P.A.; Matthews, M.V. Precursory seismic quiescence: A preliminary assessment of the hypothesis. *Pure Appl. Geophys.* **1988**, *126*, 373–406. [CrossRef]

39. Varga, P.; Grafarend, E. Distribution of the lunisolar tidal elastic stress tensor components within the Earth's mantle. *Phys. Earth Planet. Inter.* **1996**, *93*, 285–297. [CrossRef]

40. Doglioni, C.; Carminati, E.; Petricca, P.; Riguzzi, F. Normal fault earthquakes or graviquakes. *Sci. Rep.* **2015**, *5*, 1–12. [CrossRef]

41. Pétrélis, F.; Chanard, K.; Schubnel, A.; Hatano, T. Earthquake sensitivity to tides and seasons: Theoretical studies. *J. Stat. Mech. Theory Exp.* **2021**, *2021*, 023404. [CrossRef]

42. Thomas, A.M.; Bürgmann, R.; Shelly, D.R.; Beeler, N.M.; Rudolph, M.L. Tidal triggering of low frequency earthquakes near Parkfield, California: Implications for fault mechanics within the brittle-ductile transition. *J. Geophys. Res.* **2012**, *117*, 1–24. [CrossRef]
43. Stein, R.S. The role of stress transfer in earthquake occurrence. *Nature* **1999**, *402*, 605–609. [CrossRef]
44. Ellsworth, W.L.; Beroza, G.C. Seismic evidence for an earthquake nucleation phase. *Science* **1995**, *268*, 851–855. [CrossRef] [PubMed]
45. Mignan, A.; Di Giovambattista, R. Relationship between accelerating seismicity and quiescence, two precursors to large earthquakes. *Geophys. Res. Lett.* **2008**, *35*. [CrossRef]
46. Geller, R.J.; Jackson, D.D.; Kagan, Y.Y.; Mulargia, F. Earthquakes cannot be predicted. *Science* **1997**, *275*, 1616–1616. [CrossRef]

applied
sciences

Article

System-Analytical Method of Earthquake-Prone Areas Recognition

Boris A. Dzeboev [1,2,*], **Alexei D. Gvishiani** [1,3], **Sergey M. Agayan** [1], **Ivan O. Belov** [1], **Jon K. Karapetyan** [1,4], **Boris V. Dzeranov** [1,2] and **Yuliya V. Barykina** [1]

[1]   Geophysical Center of the Russian Academy of Sciences (GC RAS), 119296 Moscow, Russia;
      adg@wdcb.ru (A.D.G.); s.agayan@gcras.ru (S.M.A.); i.belov@gcras.ru (I.O.B.); jon_iges@mail.ru (J.K.K.);
      b.dzeranov@gcras.ru (B.V.D.); u.barykina@gcras.ru (Y.V.B.)
[2]   Geophysical Institute (GPI VSC RAS), Vladikavkaz Scientific Center RAS, 362002 Vladikavkaz, Russia
[3]   Schmidt Institute of Physics of the Earth of the Russian Academy of Sciences (IPE RAS),
      119296 Moscow, Russia
[4]   Institute of Geophysics and Engineering Seismology after A. Nazarov (IGES NAS RA), Gyumri 3115, Armenia
*   Correspondence: b.dzeboev@gcras.ru; Tel.: +7-495-930-05-46

Abstract: Typically, strong earthquakes do not occur over the entire territory of the seismically active region. Recognition of areas where they may occur is a critical step in seismic hazard assessment studies. For half a century, the Earthquake-Prone Areas (EPA) approach, developed by the famous Soviet academicians I.M. Gelfand and V.I. Keilis-Borok, was used to recognize areas prone to strong earthquakes. For the modern development of ideas that form the basis of the EPA method, new mathematical methods of pattern recognition are proposed. They were developed by the authors to overcome the difficulties that arise today when using the EPA approach in its classic version. So, firstly, a scheme for the recognition of high seismicity disjunctive nodes and the vicinities of axis intersections of the morphostructural lineaments was created with only one high seismicity learning class. Secondly, the system-analytical method FCAZ (Formalized Clustering and Zoning) has been developed. It uses the epicenters of fairly weak earthquakes as recognition objects. This makes it possible to develop the recognition result of areas prone to strong earthquakes after the appearance of epicenters of new weak earthquakes and, thereby, to repeatedly correct the results over time. It is shown that the creation of the FCAZ method for the first time made it possible to consider the classical problem of earthquake-prone areas recognition from the point of view of advanced systems analysis. The new mathematical recognition methods proposed in the article have made it possible to successfully identify earthquake-prone areas on the continents of North and South America, Eurasia, and in the subduction zones of the Pacific Rim.

Keywords: system-analytical method; earthquake-prone areas; pattern recognition; clustering; machine learning; earthquake catalogs; high seismicity criteria

Citation: Dzeboev, B.A.; Gvishiani, A.D.; Agayan, S.M.; Belov, I.O.; Karapetyan, J.K.; Dzeranov, B.V.; Barykina, Y.V. System-Analytical Method of Earthquake-Prone Areas Recognition. *Appl. Sci.* **2021**, *11*, 7972. https://doi.org/10.3390/app11177972

Academic Editor: Giuseppe Lacidogna

Received: 30 July 2021
Accepted: 26 August 2021
Published: 28 August 2021

1. Introduction

As a rule, strong earthquakes may not occur over the entire territory of a seismically active region. Critical objectives of the seismic hazard assessment include recognition of the areas prone to strong earthquakes. An effective instrument to accomplish this objective is pattern recognition. The fundamental possibility of employment methods and algorithms for pattern recognition to identify potentially high seismicity areas was first substantiated by remarkable mathematician I.M. Gelfand et al. in 1972 [1,2]. The developed approach was later called EPA (Earthquake-Prone Areas) [3–7].

The EPA method was developed in the fundamental papers of I.M. Gelfand and V.I. Keilis-Borok, members of the Academy of Sciences of the USSR; A.D. Gvishiani, academician of the RAS; Al.An. Soloviev, associate member of the RAS; and famous Soviet and Russian scientists, namely Sh.A. Guberman, M.P. Zhidkov, V.G. Kossobokov, A.I. Gorshkov,

V.A. Gurvich, E.Ya. Rantsman, I.M. Rotvain, etc. Prominent foreign geophysicists, seismologists, geologists, and mathematicians took an active part in developing EPA. These include F. Press and L. Knopoff, members of the United States National Academy of Sciences; Professors A. Cisternas, J. Bonnin, E. Philip, C. Weber, and J. Sallantin, French scientists; as well as M. Caputo and G. Panza, members of the National Academy of Sciences of Italy, etc. [5–7].

In the classical Gelfand–Keilis-Borok setting, the problem of strong earthquake-prone areas recognition (EPA problem) is formulated as follows. In a considered seismically active region, it is necessary to recognize the areas prone to strong earthquakes (with magnitude $M \geq M_0$, where M_0 is a given threshold). These areas are sought among the recognition objects identified in the region. As the recognition objects, morphostructural nodes or intersections of morphostructural lineaments obtained as a result of morphostructural zoning (MSZ) of the region are considered [8–12]. It is necessary to divide the set of recognition objects W into two non-intersecting classes: class B consisting of the objects whose vicinities are prone to strong earthquakes and class H, which is composed of the objects where such earthquakes cannot occur.

This classification is carried out with the employment of the pattern recognition algorithm with learning. It uses the learning set W_0, which is determined based on the information about the seismicity of the region. In turn, W_0 consists of two non-intersecting subsets B_0, containing objects that are a priori referred to as class B, and H_0, containing the representatives of class H. The result of applying the recognition algorithm is a decision function based on which an object from W can be attributed to class B or H, and the classification of the objects itself [4–6].

A detailed literary review for almost half a century of the development and application of pattern recognition algorithms to solve the recognition problem of areas prone to strong earthquakes (EPA approach), carried out by the authors of this article, is given in [5]. This paper examines the applied pattern recognition algorithms, the studied regions, and methods for assessing the reliability of the results obtained, including the theory of dynamic and limit recognition problems. This work is devoted to the presentation of modern mathematical methods developed at the Geophysical Center of the Russian Academy of Sciences, aimed at overcoming the difficulties that may arise today when using the EPA approach in its classic 50-year version. In the present paper, the problem of earthquake-prone areas recognition is considered in the classical formulation of Gelfand–Keilis-Borok.

Talking about the drawbacks of the EPA method, the following should be noted. The formation of learning material is a fundamental phase of recognition. Learning set B_0 includes the objects with known epicenters of strong earthquakes in the vicinities. It is reasonable to assume that set B_0 formed in this way is highly likely to contain no a priori errors or so few of them that they are unable to affect significantly the recognition result. It is hard to form a similar "clean" learning material of class H. The set H_0 contains either all objects not included in B_0 or objects with known earthquakes with magnitude $M < M_0 - \delta$, where $\delta > 0$ [5,13] in the vicinities.

As regards the essence of the EPA problem, which represents a threshold problem of recognition [4,14,15], learning class H_0 entails inherent potential errors. Accordingly, a low seismicity learning class is not at all a totality of benchmark objects that cannot be related to strong earthquake-prone areas. Learning sets B_0 and H_0 turn out to be disparate [16,17], and the EPA procedure ignores this fact [13].

Many years of applying the EPA method in numerous mountainous countries of the world [5] demonstrated the need to avoid learning asymmetry in recognition. Making EPA findings more reliable necessitates amending the recognition unit by adding a learning algorithm based on the only high seismicity class B_0, which includes objects with known strong earthquakes in their vicinities [13]. The development of this kind of recognition algorithm has been one of the main objectives of this study.

Another drawback of the EPA method is that the identification of recognition objects and the measurement of their geological-geophysical and geomorphologic characteristics is a time- and effort-consuming problem. That said, the possibility of using selected objects needs separate justification for every region [5]. The foregoing illustrates that the practical employment of the EPA method is still challenging to a great extent. This forced the authors to develop new state-of-the-art algorithmic systems, which enable the automation of the recognition process. The key objective of study was to set up and develop this kind of system.

2. Materials and Methods

2.1. Recognition of the Areas Prone to Strong Earthquakes with One Learning Class

In the papers [4,5], the EPA approach was further developed by creating a new algorithm called Barrier-3, employed in the recognition unit [13,16,17]. It should be especially noted that the principal difference of the Barrier-3 algorithm from the dichotomy algorithms previously used in EPA is that learning is based only on B_0 high seismicity class set.

Broadly speaking, the Barrier-3 algorithm, which learns based only on one class, is not a dichotomy algorithm. However, it can be used in the EPA. Barrier-3 also divides the territory into two non-intersecting areas that are either prone or not prone to strong earthquakes.

Similarly, to the dichotomy algorithms [5], the Barrier-3 algorithm views disjunctive nodes or axis intersections of the morphostructural lineaments as recognition objects. Such selection of objects derives from their deep tectonic connection with strong earthquakes. The confinedness of the strong earthquake epicenters to the intersections of morphostructural lineaments was statistically confirmed in the paper [18].

The objective of the Barrier-3 algorithm is to study the characteristics of learning set B_0 of the only high seismicity class and identify the objects that are "similar" to the learning objects based on the knowledge obtained. The latter are declared as high seismicity ones. Speaking the language of the theory of sets, Barrier-3 accomplishes the objective of constructing in the finite set of objects W and its subset B, broadening the only learning class B_0. For this reason, the disparity measure between two arbitrary objects is constructed for every characteristic. This allows finding and measuring the "barrier," which divides these objects as part of the considered characteristic. This assessment acts as a proximity measure for the initial set W, which enables giving exact meaning to the idea of proximity to B_0 based on a given totality of characteristics. Let us move on to the description of the mathematical construction of the Barrier-3 algorithm [13,16].

Let us assume that $\Pi = \{\pi\}$ is a finite totality of numerical characteristics of recognition objects $w \in W$, and $\pi : W \to R$, B_0 is own subset in W for learning of high seismicity class B. Based on the totality of characteristics Π of a set of objects W, it is necessary to construct a set $P_\Pi(B_0)$ that would adequately expand B_0 in the sense of requirements of the formulated problem.

The proximity of the objects w_1 and w_2 by characteristic π is "hampered" by all those objects w whose $\pi(w)$ values lie in between the values $\pi(w_1)$ and $\pi(w_2)$. They make up a barrier:

$$\mathrm{B}_\pi(w_1, w_2) = \{w \in W : min(\pi(w_1), \pi(w_2)) \leq \pi(w) \leq max(\pi(w_1), \pi(w_2))\} \qquad (1)$$

It is natural to assume that the lower the barrier $\mathrm{B}_\pi(w_1, w_2)$, the better it is for the proximity of w_1 and w_2 on W by characteristic π. This observation explains the name of the algorithm.

Let us call the ratio

$$\rho_\pi(w_1, w_2) = |\mathrm{B}_\pi(w_1, w_2)| / |W| \qquad (2)$$

disparity measure between w_1 and w_2 by characteristic π or simply barrier measure.

The Barrier-3 algorithm forms the set of $P_\Pi(v)$ objects that are close to $v \in B_0$ from W based on the characteristics Π in three phases.

Phase one: formation of set $P_\pi(v)$ of objects that are close to v in W by characteristic π using the minimality threshold $\alpha_\pi(v)$:

$$P_\pi(v) = \{w \in W : \rho_\pi(w, v) \leq \alpha_\pi(v)\} \tag{3}$$

The threshold $\alpha_\pi(v)$ performs the functions of the flexible lower boundary of the set $\{\rho_\pi(w, v),\ w \in W\}$ and can be obtained, for example, using Kolmogorov averaging with the value $s < 0$:

$$\alpha_\pi(v) = \left(\frac{\sum_{\overline{w} \in W} \rho_\pi(\overline{w}, v)^s}{|W|} \right)^{1/s} \tag{4}$$

Phase two: formation on W of the value $p_\Pi(w|v)$ showing the proximity of w to v based on all characteristics Π:

$$p_\Pi(w|v) = |\{\pi \in \Pi : w \in P_\pi(v)\}|. \tag{5}$$

Here $P_\pi(v)$ is determined by the formula (3), and the integral exponent $p_\Pi(w|v)$ introduced by the formula (5) varies from 0 to $|\Pi|$.

Phase three: formation in W of a subset of $P_\Pi(v)$ objects that are close to v based on all characteristics Π using the maximality threshold $\beta_\Pi(v)$:

$$P_\Pi(v) = \{w \in W : p_\Pi(w|v) \geq \beta_\Pi(v)\} \tag{6}$$

The threshold $\beta_\Pi(v)$ performs the functions of flexible upper boundary of the set of values $\{p_\Pi(w|v),\ w \in W\}$. Similarly, to the formula (4), $\beta_\Pi(v)$ can be constructed using Kolmogorov averaging with the value $q > 0$:

$$\beta_\Pi(v) = \left(\frac{\sum_{\overline{w} \in W} p_\Pi(\overline{w}|v)^q}{|W|} \right)^{1/q} \tag{7}$$

As a result, a sought set $P_\Pi(B_0)$ is obtained by the formula:

$$P_\Pi(B_0) = \cup_{v \in B_0} P_\Pi(v). \tag{8}$$

For the quantitative assessments of the contribution of characteristics to the formation of a sought subset of high seismicity objects, the algorithm features additional computational units. In parallel with the computation of $p_\Pi(w|v)$, a binary matrix is formed $M_\Pi(w|v)$:

$$M_\Pi(w|v)_{i,j} = \begin{cases} 1,\ w_i \in P_{\pi_j}(v) \\ 0,\ w_i \notin P_{\pi_j}(v) \end{cases},\ i = 1, \ldots, |W|,\ j = 1, \ldots, |\Pi|. \tag{9}$$

Every element of the matrix (9) determines whether or not the object $w \in W$ belongs to the set $P_{\pi_j}(v)$, $\pi_j \in \Pi$ of the objects that are close to v. String summation of the matrix $M_\Pi(w|v)$ for all $w \in P_\Pi(v)$ forms the vector $W_\Pi(v)$:

$$W_\Pi(v)_j = \sum_k M_\Pi(w|v)_{k,j},\ w_k \in P_\Pi(v),\ j = 1, \ldots, |\Pi|. \tag{10}$$

The elements of the vector (10) illustrate the contribution of characteristics $\pi_j \in \Pi$ to the formation of a subset $P_\Pi(v)$ of the objects that are close to $v \in B_0$. The quantitative assessments of the contribution of characteristics to the formation of $P_\Pi(B_0)$ are undertaken in two phases:

Element-by-element summation of all vectors $W_\Pi(v)$, $v \in B_0$ and normalization on $|B_0|$ allows obtaining the average contribution of characteristics to the recognition of a sought high seismicity subset $P_\Pi(B_0)$.

The sorting of $W_\Pi(v)$, $v \in B_0$ and the selection for each of them of three characteristics with the greatest values, followed by the summation of the number of belongings of such characteristics to the formed threes, enable assessing the contribution of such characteristics through their classification as the "strongest." This class will be called Top 3 ranking. This explains the name of the algorithm Barrier-3.

The set of recognition objects is represented as a disjoint union of $W = B \bigsqcup H$ high and low seismicity classes, $B = P_\Pi(B_0) \supseteq B_0$, and $H = W \backslash B$.

The Barrier-3 algorithm was employed as the EPA recognition block to recognize the areas prone to crustal earthquakes with $M \geq 6.0$ in the Caucasus and the Altai–Sayan–Baikal region. Here, 16 intersections of the axes of morphostructural lineaments with known epicenters of crustal earthquakes with $M \geq 6.0$, starting from 1900, in their vicinities (with a radius of 50 km in the Altai–Sayan–Baikal region and 25 km in the Caucasus) were used as the learning sets B_0 of high seismicity class in both regions.

The list of the geological-geophysical and geomorphological characteristics of considered vicinities of lineament intersections used for recognition by Barrier-3 algorithm is given in Table 1. Highlighted in bold type are 7 characteristics that were selected to be used for recognition in the Altai–Sayan–Baikal region based on the findings from the assessment of informativeness for the instance of one learning class; in italics, 11 characteristics selected for recognition in the Caucasus are given. It is noteworthy that 4 characteristics (Hmin, Top, R2, and dB) were selected for recognition using the Barrier-3 algorithm in both regions. Based on the threshold magnitude of the recognized earthquake-prone areas ($M \geq 6.0$), the circles with a radius of 25 km were selected as vicinities within which the values of characteristics were computed. For reproducibility of the result and its greater reliability, the values of characteristics of objects were computed automatically using the smart GIS developed by the Geophysical Center of the Russian Academy of Sciences (http://seismgis.gcras.ru/ access date: 30 July 2021) [19,20].

Table 1. The list of characteristics of the recognition objects (intersections of lineaments). Bold type—characteristics used by the Barrier-3 algorithm for recognition in the Altai–Sayan–Baikal region; italics—the Caucasus; bold italics—both regions.

Maximum height	*Hmax*
Minimum height	***Hmin***
The range of heights	*dH = Hmax-Hmin*
Distance between points where Hmax and Hmin are measured	l
Height gradient	dH/l
The combination of relief types	***Top***
The area of Quaternary sediments	*Q*
The highest rank of lineament	*HR*
The number of lineaments at the intersection	NL
The distance to the nearest intersection	*Rint*
Number of lineaments in the neighborhood of the intersection	*NLC*
The distance to the nearest lineament of rank I	*R1*
The distance to the nearest lineament of rank II	***R2***
The maximum value of the Bouguer anomaly	**Bmax**
The minimum value of the Bouguer anomaly	**Bmin**
The range of the Bouguer anomaly values	***dB = Bmax-Bmin***
The maximum value of magnetic anomaly	**MOmax**
The minimum value of magnetic anomaly	MOmin
The range of the magnetic anomaly values	Modif = MOmax-MOmin

2.2. Recognition of Strong Earthquake-Prone Areas Based on Identifying Dense Condensations of Point Objects

In this part of the article, let us depart from the recognition objects described above (disjunctive nodes and intersections of the axes of morphostructural lineaments) and their geological-geophysical and geomorphologic characteristics. The task considered here will be closer to reality. Namely, in addition to the reliable classification of a finite set of point objects, a formalized and reproducible transition from a recognized high seismicity set B to a real two-dimensional region in the plane with the cardinality of the continuum will be required. Strong earthquakes can occur within and cannot occur outside of this sought region. In other words, the innovation we make to the statement of the problem will be building an image of the set of recognition objects W in the studied region S as in the subset of the Euclidean plane:

$$F_\gamma : W \to F_\gamma(W) \subset S \subset R^2, \tag{11}$$

where γ is a set of free mapping parameters F_γ [21].

A sought mapping F_γ must meet the following necessary conditions:

(a) flat set $F_\gamma(B) \subset S$ is obtained for $B \subset W$, given the fixed values of free parameters γ,
(b) flat set $F_\gamma(B)$ contains high seismicity objects $w \in B$ as points in the plane, i.e., $\forall\, w \in B \Rightarrow w \in F_\gamma(B)$, and
(c) the epicenters of known strong earthquakes ($M \geq M_0$) are located inside or at the borders of zones $F_\gamma(B)$. That said, given possible errors in the identification of historic epicenters, they can be located near the borders $F_\gamma(B)$.

It is natural to view such two-dimensional sets $F_\gamma(B)$ of the cardinality of the continuum as actual flat zones within which strong earthquakes can occur. The selection of values γ of any given variant among those that meet the conditions (a), (b), and (c) is based on the system approach using control experiments.

The algorithmic system called Formalized Clustering and Zoning (FCAZ) developed by the authors [21,22] is used as sought mapping F_γ. It represents systems analysis method concerning Discrete Mathematical Analysis (DMA) [23–30].

The FCAZ method enables an effective recognition of strong earthquake-prone areas based on the cluster analysis [31] of a catalog of seismic events. It represents a consistent application of Discrete Perfect Sets (DPS) algorithms [21,32–34] and E^{2}XT [22] (Figure 1). Unlike the EPA procedure, the FCAZ systems analysis method uses neither morphostructural zoning nor dichotomy learning algorithms. It relies on the topological filtering of a finite set of epicenters of fairly weak earthquakes, which act as recognition objects.

The fundamental difference of the FCAZ systems method from the EPA procedure is that FCAZ has a formalized block (E^{2}XT algorithm) of passing from the classification of a finite set of point objects to sought flat high seismicity zones. The E^{2}XT block constructs an unambiguous mapping of a set of objects identified by the DPS algorithm in the flat zones of non-zero measure. Strong earthquakes can occur within and at the boundary of such zones. Such mapping allows for the first time to switch in the problem of earthquake-prone areas recognition from the simple pattern recognition to full-fledged systems analysis. Specifically, this makes it possible to unambiguously isolate, using sharp bound, a subsystem of recognized high seismicity zones from their non-empty complement.

The core of FCAZ is the DPS topological filtering algorithm (Figure 1) [21] isolating clusters as own subsets in a set W. This is what makes DPS different from classical clustering algorithms. It is aimed at isolating in a finite set of Euclidean space of flat regions with a given density level α.

It should be emphasized that DPS is effective for the considered problem exactly because it distinguishes between compact, connected groups of objects and their fuzzy, unstructured complement. In other words, DPS cuts out isolated objects, "attracting" the rest into dense clusters. That said, unlike classical clustering algorithms, in DPS by no means all objects end up in clusters. This is what makes DPS new and innovative as a systems analysis algorithm.

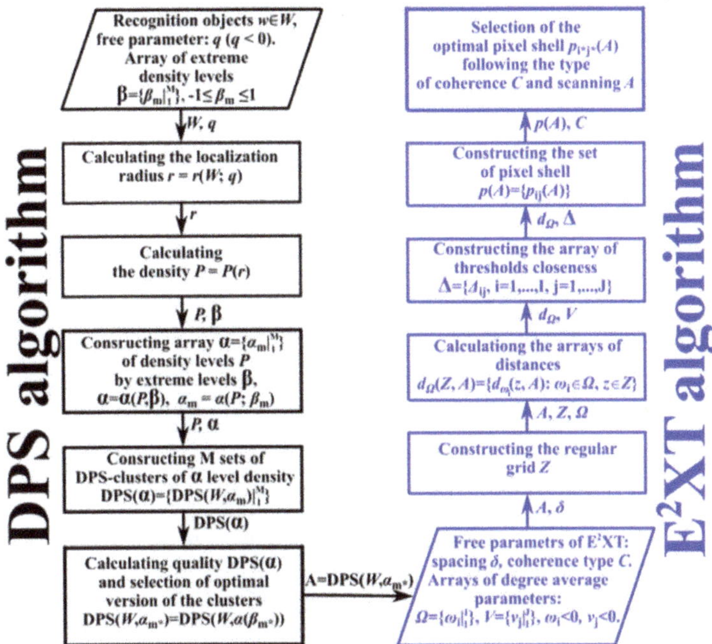

Figure 1. The flow chart of system-analytical FCAZ method. Black blocks—DPS algorithm steps; blue blocks—E^{2}XT algorithm.

The DPS algorithm has two free parameters: $q < 0$ for the calculation of localization radius $r_q(W)$, which is determined as the power mean of all nontrivial pairwise distances in a set of recognition objects W, and $\beta \in [-1,1]$ is the maximality level of the necessary density of $\alpha = \alpha(\beta, q)$ DPS clusters. The output value is a set $(\alpha(\beta, q))$ α, which is dense in each of its elements.

The result of applying the DPS algorithm is as follows:

$$\mathrm{DPS}(q, \beta) : W \to \{B_1, \ldots, B_n\}, \tag{12}$$

where parameters q and β determine a particular type of DPS clusters; $B_1, \ldots, B_n$, own connected subsets in a set of recognition objects. In other words, $B = \coprod_{i=1}^{n} B_i$, where B_i, $i = 1, \ldots, n$ are recognized DPS clusters, B, $B_i \subset W$, and $W \backslash B$ represents a significant part of the set W. It is noteworthy that if the latter is false, then the recognition result is trivial.

During the next phase, DPS clusters $B_1, \ldots, B_n$ are transformed by the E^{2}XT algorithm into flat zones of the cardinality of the continuum (i.e., mapping of F_γ is performed). During this phase, mapping is constructed:

$$\mathrm{E}^2\mathrm{XT}(\delta, C, \omega, v) : B_i \to F(B_i), \tag{13}$$

where δ is the step of the geographical grid, $\omega < 0$, $v < 0$ are free parameters, C is connection type, B_i are determined by the formula (12), $F(B_i)$, $i = 1, \ldots, n$ are sought flat zones of non-zero measure. If mapping (13) meets the above conditions (a), (b), and (c), then $F(B_i)$ there are sought areas prone to strong earthquakes.

Accordingly, the FCAZ system is a composition of two algorithms:

$$\mathrm{FCAZ}(\gamma) = \mathrm{E}^2\mathrm{XT}(\delta, C, \omega, v) \circ \mathrm{DPS}(q, \beta), \tag{14}$$

and γ is a set of free parameters of FCAZ:

$$\gamma = \{\delta, C, \omega, v\} \cup \{q, \beta\} = \{\delta, C, \omega, v, q, \beta\}. \tag{15}$$

The existence of mappings (13–15) allows considering FCAZ as a systems analysis method. FCAZ(γ), indeed, processes input data from beginning to end in a unified system. That said, the solution presented takes shape of sought two-dimensional zones and not their palliatives, representing finite sets of points in the plane [21].

The constructions of DPS and E^{2}XT algorithms feature artificial intelligence blocks that automatically select optimal values β in DPS and ω, v in E^{2}XT. This makes the result of FCAZ recognition objective and reproducible. An optimal value β enables recognizing DPS clusters for which the difference between the totality of densities (in the inherent sense of the algorithm) of objects inside the clusters and the totality of densities of objects outside of the clusters will be the greatest possible. The selected ω and v make it possible to find an optimal combination of connection and scannability of DPS clusters. A detailed description of mathematical constructions of the DPS and E^{2}XT algorithms is provided in the papers [21,22].

FCAZ delivers a system approach to studying strong earthquake-prone areas. Its characteristic property lies in the fact that the recognition of sought zones in the regions of the globe that differ in structure relies on universal facts and methods, which enable a uniform approach toward solving the entire class of such problems. The statement of the problem and the process of its solving represent a unified system that is sufficiently invariable relative to geological structure, the selection of threshold magnitudes of sought strong earthquakes, objects, etc. That said, the FCAZ applicability condition is the state of seismological and geological-geophysical exploration of regions, which manifests itself in the high quality of earthquake catalog.

The next parts of the paper will demonstrate how FCAZ allows performing reliable recognition of the strongest ($M \geq 7.75$), strong ($M \geq 6.0$), and significant earthquake-prone areas across different mountain countries of the world. The reliability of obtained results was assessed with the help of control experiments and through comparison with high seismicity zones recognized by EPA approach. The division of earthquakes into the strongest, strong, and significant ones was made by the authors solely to simplify the description of the results obtained.

It should be noted that previously the recognition of high seismicity zones was performed just for one fixed magnitude threshold M_0. In the present paper, a method for the successive recognition of the areas prone to earthquakes for different threshold magnitudes in the same region is proposed. It is based on the repeated application of FCAZ method to a set of recognition objects, which is successively narrowed down by way of DPS clustering. The new method was called Successive Formalized Clustering and Zoning and is abbreviated as SFCAZ [35]. Accordingly, the classical EPA problem formulated at the beginning of the article is for the first time expanded to a more systemic complicated problem of the successive recognition of the areas prone to earthquakes in the same region for several threshold magnitudes. The mathematical construction of the SFCAZ method is described in detail in the paper [35].

3. Results

3.1. Variable EPA Method

Figure 2 shows the MSZ map of Altai–Sayan–Baikal region. The result of earthquake-prone areas recognition with $M \geq 6.0$ by Barrier-3 algorithm is shown in Figure 2 using ellipses with blue boundaries. Upon the completion of recognition, 32 out of 97 objects are identified as high seismicity class B. The totality of vicinities of these objects (circles used to compute the values of characteristics) [5] determines the areas prone to earthquakes with $M \geq 6.0$.

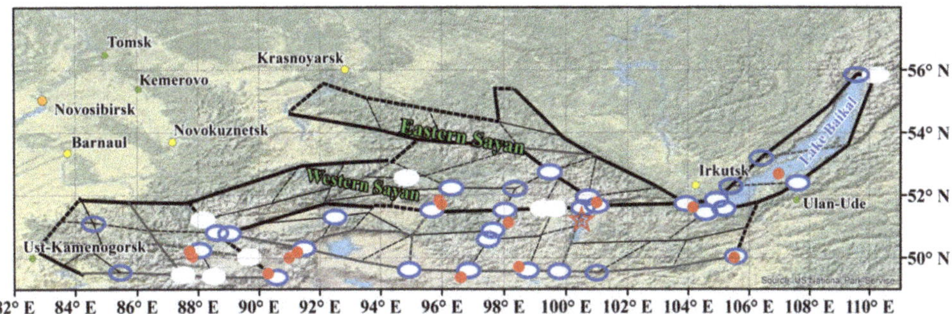

Figure 2. The morphostructural zoning map of the Altai–Sayan–Baikal region. Thick black lines—rank I lineaments; medium gray lines—rank II; thin black lines—rank III; solid lines—longitudinal lineaments; dotted line—transverse ones [36,37], areas prone to earthquakes with $M \geq 6.0$ (ellipses with blue boundaries—Barrier-3 [16], white ellipses—Cora-3 [37], white ellipses with blue boundaries—both the algorithms). Red circles designate the epicenters of crustal earthquakes with $M \geq 6.0$ (1900–2012) used to form the learning set B_0, red star refers to the epicenter of crustal earthquake (11 January 2021 with $M = 6.7$) that occurred after the completion of recognition.

Figure 3a shows a bar diagram characterizing the medium contribution of characteristics in recognition using the Barrier-3 algorithm of a sought high seismicity set of objects. Figure 3b shows the contribution of characteristics expressed through attribution to the Top 3 rankings. The y axis in Figure 3a shows the average number of "attributions" of characteristics in the recognition of a set $P_\Pi(B_0)$ (see above), in Figure 3b, the number of attributions to the three "strongest" characteristics (Top 3 ranking).

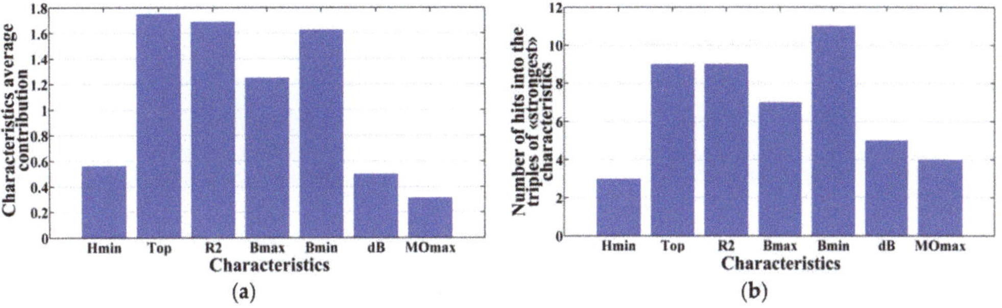

Figure 3. The contribution of characteristics to recognition using the Barrier-3 algorithm of a high seismicity set of objects in the Altai–Sayan–Baikal region: (**a**) average contribution of characteristics; (**b**) the contribution of characteristics expressed through their attribution to the "strongest" threes.

It can be seen from Figure 3 that in the recognition of strong earthquake-prone areas in the Altai–Sayan–Baikal region using the Barrier-3 algorithm, the most significant characteristics include gravity anomalies (Bmax and Bmin), a combination of relief types (Top) and the distance to the nearest lineament of rank II (R2). The intersections of lineaments classified as high seismicity ones against the background of the entire set of objects in their vicinities are characterized by low values of gravity anomalies (mostly Bmax ≤ -160 mGal and Bmin ≤ -220 mGal) and contrasting combinations of relief types—mountains/foothill and mountains/mountains. These are characterized by high values of magnetic anomaly (MOmax), the concentration of dB around 60 mGal and 120 mGal, and the concentration of Hmin less than -1000 m and more than 1000 m.

It can be seen from Figure 3 that lithospheric magnetic field anomalies contribute to the result of recognition in the Altai–Sayan–Baikal region. It can thus be concluded that the vicinities of high seismicity intersections of lineaments are characterized by a high degree of tectonic breaks, the existing deep density heterogeneity, as well as specific structure and composition of the Earth's crust. It would be natural to interpret these signs as the criteria of high seismicity in the studied region.

A comparative analysis of results, obtained independently using the Barrier-3 algorithm and the Cora-3 dichotomy algorithm, which is most common in the EPA [38,39], shows that they are well aligned with each other (Figure 2). The Barrier-3 algorithm recognized as high seismicity ones 32 intersections of morphostructural lineaments and Cora-3 33 intersections [36,37]. That said, 25 objects were attributed by both algorithms to class B.

The Barrier-3 algorithm classified as hazardous 6 out of 51 objects of the learning set of a low seismicity class of dichotomy; Cora-3 3 out of 51; both algorithms, 2 intersections. Consequently, 44 learning objects of class H were recognized by both algorithms as non-hazardous for magnitude $M \geq 6.0$. It means that the key differences in the classification belongs to a set of objects initially not attributed to learning sets (20 objects are classified identically by the algorithms and 10 are classified differently). It should be noted that the epicenters of earthquakes with $M \geq 6.0$, used to form the learning set B_0 (red circles in Figure 1) are located strictly within the vicinities of objects classified by both algorithms as high seismicity ones [16,37].

It is noteworthy that the Barrier-3 algorithm is structured in such a way that learning objects in the final classification always belong to class B. In turn, in recognition using dichotomy algorithms (particularly Cora-3 algorithm), learning objects are broadly speaking, not obliged to retain their attribution to the relevant class [4,7].

The red star in Figure 2 shows the epicenter of the crustal earthquake, which occurred on 11 January 2021, with $M = 6.7$. This earthquake occurred after the completion of the independent recognitions described herein using the Barrier-3 and Cora-3 algorithms, thus representing the material for a pure examination for them. It can be seen from Figure 2 that the epicenter is located outside of the vicinities (with a radius of 25 km) of the intersections of lineaments recognized as high seismicity ones by both algorithms. At the same time, it is located 42 km away from the nearest recognition object attributed to class B by both algorithms. It was believed in the formation of learning material that the epicenter is confined to the intersection if it is located at a distance of no more than 50 km. Accordingly, the epicenter of the earthquake that occurred on 11 January 2021, is confined to the intersection of lineaments attributed to class B by both algorithms, yet is located outside of the 25 km of vicinities used to compute the values of its characteristics.

The recognition results of the areas prone to earthquakes with $M \geq 6.0$ in the Altai–Sayan–Baikal region, obtained using the Barrier-3 (one learning class) [16] and Cora-3 algorithms (two learning classes) [36,37], are well aligned with each other. On the one hand, this evidences the reliability of both results since recognition was performed independently. On the other hand, in the interpretation of differences (15.5% of the total number of objects) in the outcomes, preference should be given to the classification using the Barrier-3 algorithm since it was performed with learning containing no protentional errors.

The recognition result of the areas prone to strong earthquakes in the Caucasus, obtained using the Barrier-3 algorithm as the EPA recognition block, is shown in Figure 4 as ellipses with blue boundaries. Barrier-3 attributed 108 out of 237 intersections of lineaments to high seismicity class B.

Figure 5 shows the bar diagrams demonstrating the contribution of object characteristics to the recognition, by Barrier-3 algorithm, of the intersections of lineaments, in whose vicinities strong earthquakes can occur in the Caucasus. As it can be seen, the greatest contribution is made by the characteristics that are responsible for relief heights (Hmax and Hmin), the area of quaternary rocks (Q), the highest rank of lineament (HR), the number of lineaments in the vicinities (NLC), and the distances to the nearest lineaments of ranks I (R1) and II (R2).

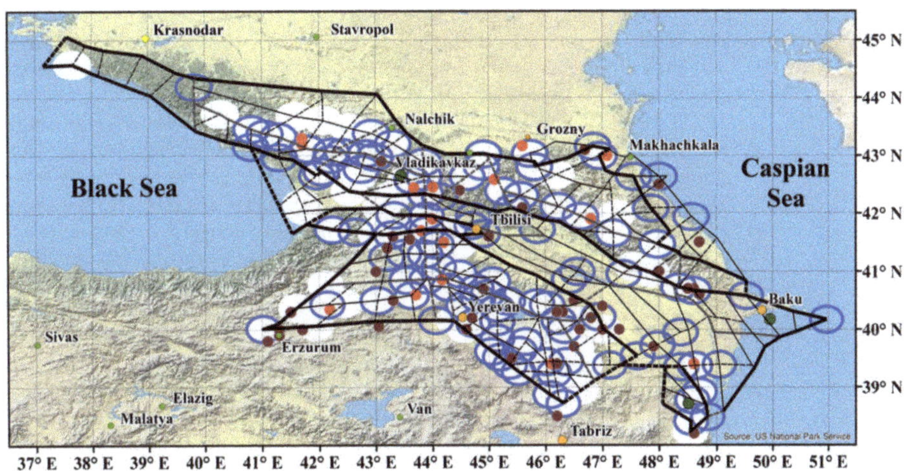

Figure 4. The morphostructural zoning map of the Caucasus (legends as in Figure 2) [40], areas prone to earthquakes with $M \geq 6.0$ (ellipses with blue boundaries—Barrier-3 [17], white ellipses—Cora-3 [41], white ellipses with blue boundaries—both algorithms) and the epicenters of earthquakes with $M \geq 6.0$ (brown circles—before 1900, red circles—in the period from 1900 to 1992 (used to form the learning set B_0), dark green circles—since 1993 (material for a pure exam)) [42].

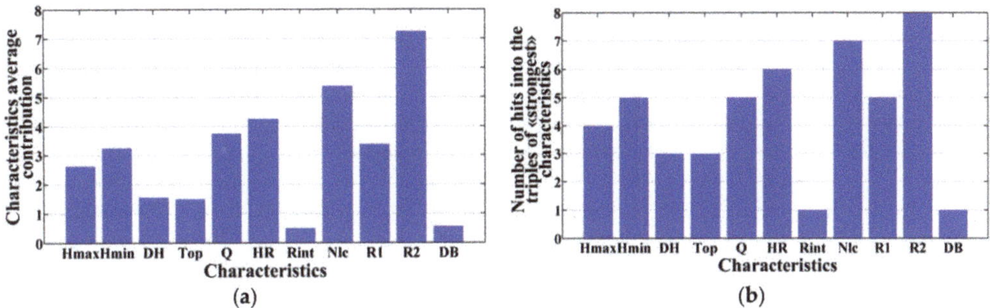

Figure 5. The contribution of characteristics to the recognition, using the "Barrier-3" algorithm, of a high seismicity set of objects in the Caucasus: (**a**) average contribution of characteristics; (**b**) the contribution of characteristics expressed through their attribution to the "strongest" threes.

In the Caucasus, the intersections of lineaments recognized as hazardous ones for $M \geq 6.0$ against the background of the entire set of recognition objects in their vicinities are characterized by high values of the maximum and minimum heights (Hmax $\geq$ 2500 m and Hmin $\geq$ 600 m) and a small area of quaternary rocks (Q $\leq$ 30%). They are made up of three or more lineaments of ranks II or III (NLC $\geq$ 3, HR = 2 or HR = 3, R2 $\leq$ 30 km) and located at a relatively short distance away from the lineaments of rank I (0 < R1 $\leq$ 50 km).

The joint analysis of Figures 3 and 5 shows that in both regions for the Barrier-3 algorithm, a significant contribution to the formation of a high seismicity set of objects is made by the distance to the nearest lineament of rank II. Namely, the characteristic R2 is important for the Barrier-3 algorithm and invariable relative to the selection out of two regions considered.

Figure 4 illustrates a comparison of the classification of lineament intersections in the Caucasus, obtained with the help of Barrier-3 and Cora-3 algorithms. The first one recognized as high seismicity ones 108 intersections of lineaments [17]; the second one, 107 [41], both algorithms simultaneously recognized 73. The Barrier-3 algorithm classified as hazardous 24 out of 71 objects of the learning set of low seismicity class from Cora-3; Cora-

3, 22; both algorithms simultaneously, 16. In total, 41 learning objects of a low seismicity class were recognized by both algorithms as non-hazardous. Out of the intersections set not initially classified as the learning sets of Cora-3, the algorithms classified identically 95 objects and classified differently 55 objects [17,41].

It can be seen from Figure 4 that the objects located on the longitudinal lineaments of rank II and classified by both algorithms as high seismicity ones make up extensive zones along the axis of the Main Ridge in the Central and Southeastern Segments of the Greater Caucasus. A good coincidence of recognition results can be seen in the eastern sector of the Lesser Caucasus and the Armenian Volcanic Plateau [43]. A totality of the objects located on the transverse lineaments of rank II and attributed by the Barrier-3 and Cora-3 algorithms to class B make up an extensive submeridional zone within the Trans-Caucasian Transverse Elevation, combining the areas prone to strong earthquakes in the Greater and Lesser Caucasus. A fairly good alignment of high seismicity areas can also be seen near the Talysh mountains. It is noteworthy that most earthquakes known in the Caucasus with $M \geq 6.0$ occurred in the vicinities of the objects making up the zones described above.

The analysis of Figure 4 showed that all 17 epicenters of earthquakes with $M \geq 6.0$ (red circles), which formed the learning set of high seismicity class of both algorithms, are located inside the B zones recognized by both algorithms. Out of 42 epicenters of strong earthquakes, which occurred before 1900 (brown circles), 7 and 8 epicenters, respectively, are located outside of the zones recognized by the Barrier-3 and Cora-3 algorithms. Half of them are located within a short distance from the potentially high seismicity areas recognized by the algorithms.

Dark green circles in Figure 4 refer to the epicenters of strong earthquakes, which have occurred in the Caucasus since 1993. Information about them has not been used, in any manner whatsoever, in the formation of learning sets; thus these earthquakes represent material for a pure examination. Two of the three epicenters are located strictly within the high seismicity zones recognized by both algorithms. The latter represents a significant argument in favor of the reliability of the result demonstrated by Figure 4.

The replacement of a dichotomy algorithm with the original Barrier-3 algorithm, undertaken in this paper, is an attempt to open a new page in the development of the EPA approach. As shown above, the Barrier-3 algorithm proved itself to be good in the recognition of strong earthquake-prone areas with one learning class in the Caucasus and the Altai–Sayan–Baikal region. This fact strengthens the assumptions that the approach toward the recognition of potentially high seismicity zones based on the only high seismicity learning class through its expansion is adequate to the classical setting of the EPA problem.

The positive variants for recognition obtained using the Barrier-3 and Cora-3 algorithms make them control experiments for each other. Due to the relative proximity of results, these control experiments should be recognized as successful. This enhances the assessment of the reliability of the above results.

The studied regions serve as a basis for the proposed joint interpretation of the strong earthquake-prone areas recognized for one and two learning classes. The interpretation relies on the composition of unclear set construction [44] and the results obtained independently using the Barrier-3 algorithm and the Cora-3 dichotomy [45].

Let W still represent a set of intersections of lineaments, and a fuzzy set of high seismicity objects is defined as a set of pairs:

$$B = \{w, \ \mu_B(w)|w \in W\}. \tag{16}$$

That said, membership function $\mu_B(w)$ is:

$$\mu_B(w) = \mu_{B_1, B_2}(w) = \begin{cases} 1, \ w \in B_1 \cap B_2 \\ 0.5, \ w \in B_1 \Delta B_2 \quad = (B_1 \cup B_2) \backslash (B_1 \cap B_2), \\ 0, \ w \notin B_1 \cup B_2 \end{cases} \tag{17}$$

where B_1 and B_2 are the sets of objects recognized as high seismicity ones by the Barrier-3 and Cora-3 algorithms, respectively. Then high seismicity objects in the integral result are the intersections for which $\mu_B(w) > 0$.

Figure 6 provides the example of an interpretation of results for strong earthquake-prone areas recognition with $M \geq 6.0$ in the Caucasus and the Altai–Sayan–Baikal region using a fuzzy set construction (16–17). In the Altai–Sayan–Baikal region in line with the obtained independent results of recognition (Figure 2), all considered epicenters of strong earthquakes are located in the vicinities of objects attributed to class B by both algorithms. Whether or not the epicenter of the 2021 earthquake should be treated as a "missed target" error, the number of missed recognition targets in cases where a fuzzy function is used (Figure 6a) is the same for each algorithm (Figure 2). In this case, recognition using the Formulas (16) and (17) only increases the number of sought high seismicity objects, where strong earthquakes have not been recorded until the present.

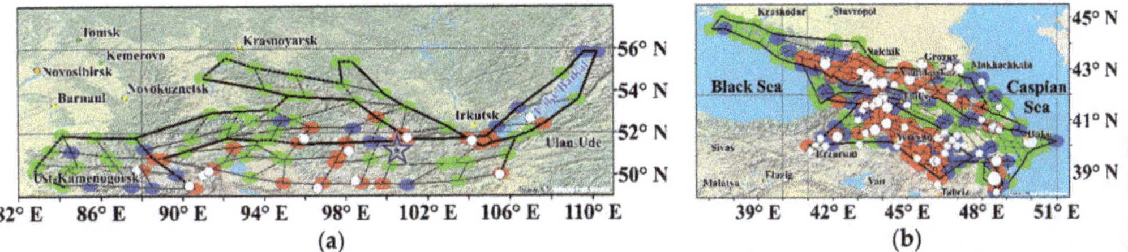

Figure 6. Presentation of a joint result of earthquake-prone areas recognition with $M \geq 6.0$ by the Barrier-3 and Cora-3 algorithms as a fuzzy set of vicinities of the intersections of lineaments: (**a**) the Altai–Sayan–Baikal region (white circles—epicenters of earthquakes with $M \geq 6.0$ (1900–2012); blue star—epicenter of the earthquake, which occurred on 11 January 2021); (**b**) the Caucasus (white circles—epicenters of earthquakes with $M \geq 6.0$; minor ones—before 1900; medium ones—1900–1992; major ones—after 1992). Highlighted in red are the vicinities of intersections of lineaments with membership function to the high seismicity set $\mu = 1$; in blue, $\mu = 0.5$; in green, $\mu = 0$. The function μ is determined by the formula (17).

The situation in the Caucasus is different. The fuzzy function approach a priori improves the quality of the result here. In Figure 4, out of 62 epicenters of the considered earthquakes with $M \geq 6.0$, 8, and 9 epicenters, respectively, lie outside of the high seismicity areas recognized by the Barrier-3 and Cora-3 algorithms. That said, as few as 4 epicenters are located outside of the zones identified as high seismicity ones (red and blue ellipses in Figure 6b) based on the Formulas (16) and (17).

The integral result (Figure 6) identifies 41.2% of objects in the Altai–Sayan–Baikal region and 59.9% in the Caucasus as high seismicity ones. That said, for the studied EPA problem, the result is typically treated as nontrivial if not more than 60% of objects are classified as high seismicity ones [4]. The recognition obtained based on the Formulas (16) and (17) meets this condition for both regions. At the same time, this allows obtaining a new nontrivial result for both regions and halving the number of missed targets in the Caucasus.

The improvement of recognition result when construction (16–17) is used derives from the fact that the employment of fuzzy mathematics enables integrating the criteria of two independent recognitions performed by the Barrier-3 and Cora-3 algorithms. This allows, to some extent, compensating incomplete and sometimes defective input data [45].

3.2. FCAZ Recognition of the Strongest Earthquake-Prone Areas

Three regions of the Pacific Seismic Belt are considered. Within their limits by the FCAZ method the areas prone to the strongest earthquakes with $M \geq 7.75$ are recognized. The epicenters of earthquakes with the focal depth of up to 70 km from the ANSS catalog (1963–2013, the mountain belt of the South American Andes), the earthquakes catalog of Kamchatka and the Commander Islands (1962–2015, the coast of the Kamchatka Peninsula), and the catalog of the Kuril–Okhotsk region (1962–2009, the coast of the Kuril Islands) are used as recognition objects. To select the magnitude threshold M_R, starting from which the epicenters were used as recognition objects, completeness magnitude M_c was assessed in the catalogs [46–48]. Taking into account M_c assessment, it was decided to use as FCAZ recognition objects in the Andes the earthquake epicenters with $M \geq M_R = 4.5$ (16,556 epicenters) [21]; in Kamchatka, $M \geq M_R = 3.5$ (44,113 epicenters) [49–51]; in the Kuril Islands, $M \geq M_R = 4.2$ (11,725 epicenters). In Figure 7a, the totality of blue and green colors shows recognition objects in the mountain belt of the South American Andes.

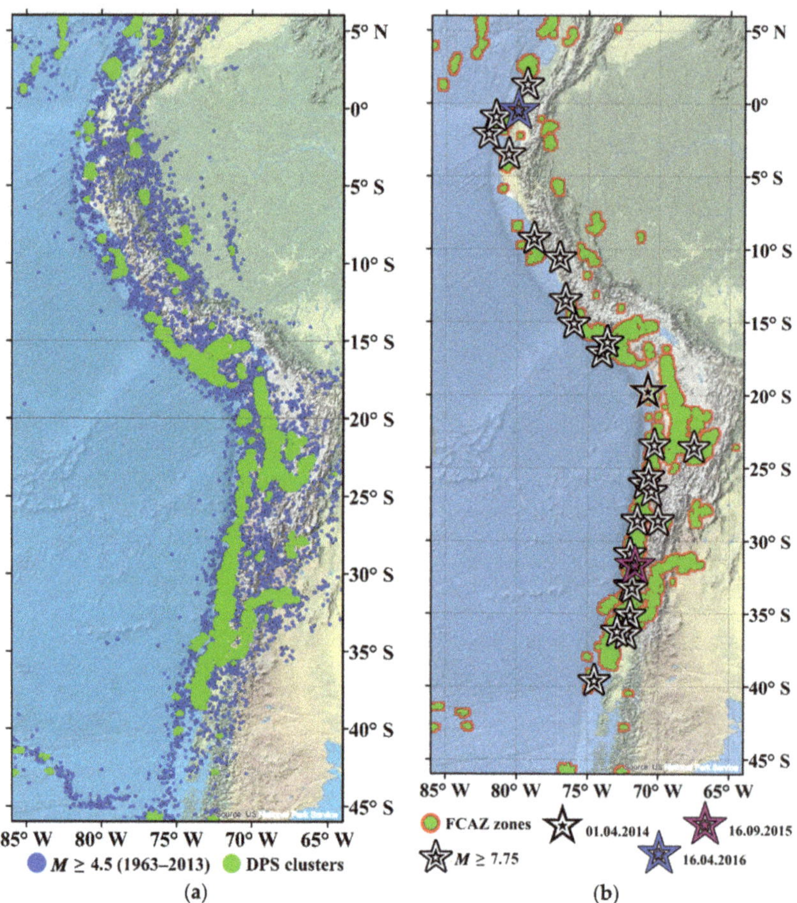

Figure 7. Mountain belt of the Andes: (**a**) FCAZ recognition objects—the epicenters of earthquakes with $M \geq 4.5$ and recognized DPS clusters; (**b**) FCAZ zones prone to earthquakes with $M \geq 7.75$ and the epicenters of earthquakes with $M \geq 7.75$.

The lists of the strongest crustal earthquakes, beginning in 1900, have been formed based on the above-listed instrumental catalogs, EPA recognition works, and the catalog of strong earthquakes in the USSR from ancient times to 1975 [52]. As a result, the catalog of the strongest earthquakes of the mountain belt of the Andes contains 24 events for the period of 1900–2013; the catalog of Kamchatka, 8 (1900–2015); and that of the Kuril Islands, 11 (1900–2009). The epicenters of earthquakes with $M \geq 7.75$ are shown in Figures 7b and 8.

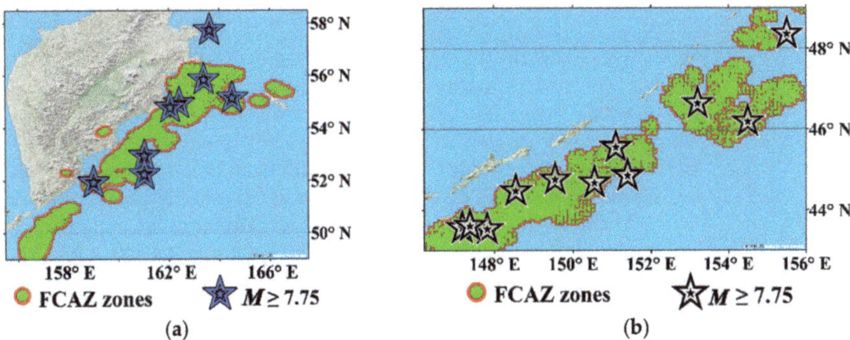

Figure 8. FCAZ zones prone to earthquakes with $M \geq 7.75$ and the epicenters of earthquakes with $M \geq 7.75$: Pacific Coast (**a**) of the Kamchatka Peninsula; (**b**) of the Kuril Islands.

The DPS clustering of the epicenters of earthquakes, which represent FCAZ recognition objects, was performed as follows. Initially, the DPS algorithm was employed with density level $\alpha_1(\beta_1)$. The obtained dense set of objects $W_1(\alpha_1(\beta_1))$ was excluded from further consideration and the algorithm was applied for the second time to the remaining subset with density level $\alpha_2(\beta_2)$. This allowed obtaining new DPS clusters $W_2(\alpha_2(\beta_2))$, where $W_2 = W \backslash W_1(\alpha_1(\beta_1))$. Subsequent iterations were performed similarly. All connected components forming part of $W_1(\alpha_1(\beta_1)) \cup W_2(\alpha_2(\beta_2)) \cup \ldots \cup W_k(\alpha_k(\beta_k))$ were declared as sought DPS clusters.

Four iterations of DPS clustering were performed in the mountain belt of the South American Andes; two iterations, in Kamchatka; and three, in the Kuril Islands. The optimal values of the β parameter—the maximality level of density of DPS clusters—were computed automatically using the artificial intelligence block. It should be noted that 67% of recognition objects in the mountain belt of the Andes were included in the recognized DPS clusters; 73.3%, in Kamchatka; 77.5%, in the Kuril Islands. DPS clusters are highlighted in green in Figures 7 and 8.

In each of the three regions, the E^2XT algorithm was applied to DPS clusters. The optimal values of its input parameters ω and v were computed using the artificial intelligence block. That said, a regular geographical graticule and connection type C_8 was used. In Figures 7b and 8, the totality of green and red colors shows mapped FCAZ zones.

Figures 7b and 8 show that FCAZ zones are well aligned with the location of the epicenters of the known strongest earthquakes. Out of 24 earthquakes with $M \geq 7.75$ in the mountain belt of the Andes, only one epicenter (4.2%) is located outside of FCAZ zones (Figure 7b) and creates a missed target error. This is an epicenter of the earthquake which occurred on 24 May 1940, more than 20 years before the commencement of systemic instrumental seismological observations in the region. Accordingly, the location of the epicenter can be distorted and this only error can be irrelevant.

Out of the eight strongest earthquakes considered in the Pacific Coast of the Kamchatka Peninsula, the epicenter of just one (12.5%) does not belong to the recognized FCAZ zones (Figure 8a). This is an epicenter of the Ozernovskiy earthquake with $M = 7.75$ in Koryakia, which occurred on 22 November 1969, in the north of the considered region.

In 2006 the Olyutorskoye earthquake occurred in Koryakia to the north of the border of the considered region, its magnitude was similar to the Ozernovskoye (Figure 8a). The missed target error of the Ozernovskoye earthquake and non-inclusion of the Olyutorskoye earthquake area in the considered region is caused by the fact that their epicenters are located outside of today's subduction zone. The conditions for the occurrence of these earthquakes outside of the subduction zone are dramatically different from the remaining considered strongest earthquakes in the region. This is also justified by the fact that the epicenters of both earthquakes lie outside of the territory in respect of which work is underway to make a long-term forecast of the strongest earthquakes using the method of Academician of RAS S.A. Fedotov [53]. Accordingly, the epicenter of the Ozernovskoye earthquake is possibly not a missed target error.

On the Pacific Coast of the Kuril Islands (Figure 8b), the epicenter of just one (9%) out of 11 known earthquakes with $M \geq 7.75$ is a missed target error (the earthquake dated 1 May 1915, with $M = 8.3$). Let us note here that the identification of the areas prone to earthquakes on the Pacific Coast of the Kuril Islands using pattern recognition methods has not been previously performed. It is undertaken in this paper for the first time.

It is noteworthy that the FCAZ zones recognized in the mountain belt of the South American Andes contain 69% of earthquake epicenters with $M \geq 5.0$ from among those present in the instrumental catalog used for recognition purposes. That said, they occupy approximately half of the area of the seismically active mountain belt of the Andes and the active subduction zone. FCAZ zones on the Kamchatka coast contain 73% of the epicenters of earthquakes, with $M \geq 4.0$ among those present in the instrumental catalog and occupying 40% of the area of seismically active Kuril-Kamchatka and Aleutian Arcs falling within the boundaries of the considered region. On the coast of the Kuril Islands, FCAZ zones contain 81% of the earthquake epicenters with $M \geq 5.0$ among those present in the catalog. The aforesaid allows interpreting, with a high degree of reliability, the recognized FCAZ zones (Figures 7b and 8) as the areas prone to earthquakes with $M \geq 7.75$ in the mountain belt of the Andes and on the Pacific Coast of the Kamchatka Peninsula and the Kuril Islands.

The recognized zones prone to the strongest earthquakes in Kamchatka and on the Kuril Islands are well aligned with the results of a long-term seismic forecast for IX 2013–VIII 2018, using the method of Academician of RAS S.A. Fedotov. In [53], earthquakes with $M = 5.7$–7.2 were expected throughout the Pacific Coast of Kamchatka with a varying probability level. That said, during the above-mentioned time interval, earthquakes with $M \geq 7.7$ were expected in the coastal zone of the Avacha Bay and near the shores of Southern Kamchatka. Fairly big FCAZ zones are situated in these areas as well (Figure 8a).

The best justification for the reliability of recognition results is a pure experiment, i.e., the analysis of the alignment of FCAZ zones and the location of the epicenters of earthquakes (with $M \geq M_0$) that occurred after the end of the instrumental catalog used for recognition purposes. For instance, three earthquakes with $M \geq 7.75$ occurred in the mountain belt of the South American Andes after 2013: on 1 April 2014, with $M = 8.2$ (northwest of the Chili coast), on 16 September 2015, with $M = 8.3$ (Chili coast), and on 16 April 2016, with $M = 7.8$ (Ecuador). Information about these strongest earthquakes was not used for recognition purposes in any manner whatsoever.

The epicenters of earthquakes of 2014, 2015, and 2016 are shown in Figure 7b using black, purple, and blue stars, respectively. The first two epicenters are located strictly inside the FCAZ zones. The third one is a short distance away from the boundaries of the recognized zones. This allowed obtaining an argument in favor of the reliability of the completed FCAZ recognition, both weighty and independent from research results.

Summarizing the results obtained, an important achievement should be noted. For the first time, the strongest earthquake-prone areas were successfully recognized based on the objective classification without involving morphostructural zoning and the formation of learning sets. That said, the results are generally well aligned with those previously obtained independently using the EPA method (for details, see below). Accordingly, it is shown that FCAZ method is applicable to the system observation of regions with a very high seismicity level.

3.3. FCAZ Recognition of the Areas Prone to Strong and Significant Earthquakes for One and Several Threshold Magnitudes

The regions with a lower seismicity level than in the previous section of the article: California, the Altai–Sayan region and the Baikal–Transbaikal region, the Caucasus, as well as the Crimean Peninsula, and the northwestern Caucasus are considered. The sets of recognition objects were formed based on the epicenters of crustal earthquakes from the following catalogs: ANSS (1960−2012, California), Earthquakes in the USSR and Earthquakes of Northern Eurasia (1962–2008, the Caucasus; 1962–2008, Crimea and northwestern Caucasus; 1962–2009, the Altai–Sayan region; and 1962–2010, the Baikal–Transbaikal region). Based on the assessment of M_c, it was decided to use the epicenters of earthquakes with $M \geq M_R = 3.0$ (31,874 epicenters) as recognition objects in California [54–56]; in the Altai–Sayan region, $M \geq M_R = 2.8$ (3647 epicenters) [57]; in the Caucasus, $M \geq M_R = 3.0$ (6980 epicenters) [21,22,58,59]; in Crimea and northwestern Caucasus, $M \geq M_R = 2.0$ (2398 epicenters) [60,61]; and in the Baikal–Transbaikal region, $M \geq M_R = 2.7$ (11,297 epicenters) [35].

In Figure 9, the totality of green and red colors shows the recognized FCAZ zones prone to strong earthquakes in California ($M \geq 6.5$) and significant earthquakes in the Altai–Sayan region ($M \geq 5.5$), in the Caucasus ($M \geq 5.0$), and in the Crimean Peninsula and northwestern Caucasus ($M \geq 4.5$).

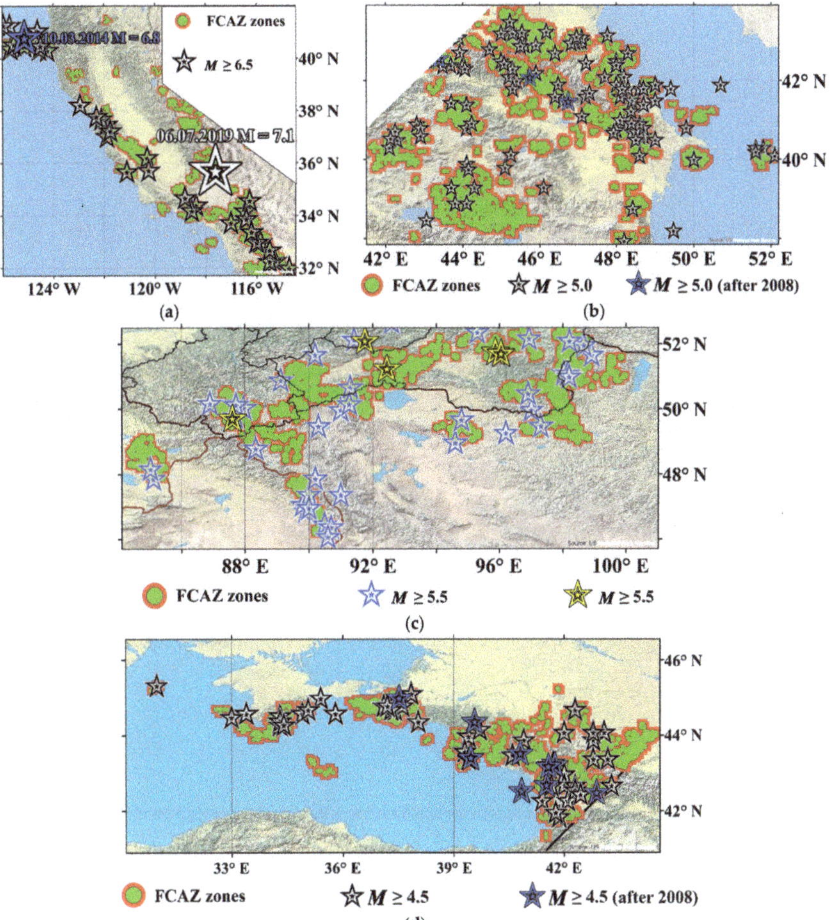

Figure 9. FCAZ zones prone to earthquakes: (**a**) California, $M \geq 6.5$ (black stars—epicenters of earthquakes with $M \geq 6.5$ from 1836 through 2010; blue and white stars—$M \geq 6.5$, since 2014); (**b**) the Caucasus, $M \geq 5.0$ (black stars–$M \geq 5.0$ for the period of 650–2008; blue stars—$M \geq 5.0$, since 2009); (**c**) the Altai–Sayan region, $M \geq 5.5$ (blue stars—$M \geq 5.5$ for the period of 1902–2008; yellow stars—$M \geq 5.5$, since 2011); (**d**) Crimea and the northwestern Caucasus, $M \geq 4.5$ (black stars—$M \geq 4.5$ for the period of 1900–2008; blue stars—$M \geq 4.5$, since 2009).

At first sight, out of 33 strong earthquakes with $M \geq 6.5$ (1836–2010) in California, the epicenters of 5 (15%) do not fall within the FCAZ zones (Figure 9a). It should be noted that three of them are located offshore in the Pacific Ocean at a great distance from the coast and are thus not caused by the tectonics of the studied region. Two other earthquakes occurred in 1857 and 1906 a long time before the commencement of systematized instrumental observations in the region. If thus these special cases are excluded from consideration, we will see that, in fact, the result contains no missed target errors [54].

It should be noted that in California, FCAZ zones contain 83% events with $M \geq 4.5$ among those present in the instrumental catalog. After the end of the catalog used to select recognition objects, two strong earthquakes occurred (Figure 9a). The epicenters of both lie strictly within the FCAZ zones. Accordingly, we have grounds to believe that recognition resulted in building sought areas prone to strong earthquakes with $M \geq 6.5$ in California.

In the Altai–Sayan region, out of 48 (1902–2008) significant earthquakes with $M \geq 5.5$, only 7 epicenters (15%) are located outside of the recognized FCAZ zones (Figure 9c). It should be noted that 6 of them occurred before the commencement of active seismological observations. That said, 3 epicenters are situated in Mongolia, 2 are located in the south of the Krasnoyarsk region, where a small number of seismic stations now function.

FCAZ zones contain 67% of the earthquake epicenters with $M \geq 4.0$ from among those present in the catalog. After the end of the used instrumental catalog, five significant earthquakes occurred in the region (Figure 9c). Of them, four epicenters are located strictly inside FCAZ zones. Summing it up, the totality of provided arguments allows stating that the results of FCAZ recognition in the Altai–Sayan region shown in Figure 9c have a high degree of reliability [57].

In FCAZ research in the Altai–Sayan region, an attempt was also made to recognize zones with the lowest possible number of missed target errors. For this purpose, localization radius $r_q(W)$ was varied in DPS clustering through changes in a preset interval of values of parameter q. Of all obtained recognition variants, an optimal one was selected, i.e., having the lowest number of omissions of significant earthquake epicenters. This variant had two missed targets less than the main variant of FCAZ zones (Figure 9c).

The epicenters of these two significant earthquakes, which make up the difference in the number of missed target errors, are located within and at the border of Mongolia. That said, the resulting area of optimal zones was 1.5 times higher than in the main recognition variant (Figure 9c). Accordingly, in the case of optimal FCAZ zones, the number of false alarms grows inevitably, adversely affecting the reliability of recognition. In this regard, the final choice was made in favor of the main recognition variant (Figure 9c) [57].

In the Caucasus, out of 106 (650–2008) significant earthquakes with $M \geq 5.0$, the epicenters of 8 (7.5%) are located outside of FCAZ zones (Figure 9b). Explaining these missed targets, note that three earthquakes occurred in 957, 1250, and 1667 long before the commencement of systemized instrumental observations in the region. Three more unrecognized epicenters are situated at a great distance from seismic networks based on which the catalogs used for recognition were created. This casts doubt on the fact that these earthquakes are real missed targets. Accordingly, certain missed targets are two epicenters of significant earthquakes.

It should be noted that FCAZ zones contain 68% of the epicenters of earthquakes with $M \geq 4.0$ among those present in the instrumental catalog. The epicenters of all three significant earthquakes which occurred after the end of the used instrumental catalog are located strictly inside of FCAZ zones (Figure 9b). This is an argument in favor of the reliability of the FCAZ recognition results [22].

It should be noted that the subregion in the northwestern part of the Caucasus (white triangle in Figure 9b) was excluded from consideration due to the lack of earthquake epicenters representing recognition objects in this region. FCAZ recognition in this subregion turns out to be impossible. This area forms part of the united region Crimea—northwestern Caucasus, in which FCAZ recognition was performed for $M_0 = 4.5$ [21].

The number of earthquakes with $M \geq M_0$ must be sufficient to assess the level of their alignment with the recognized FCAZ zones. In the instrumental catalog of earthquakes of the Crimea and northwestern Caucasus region, there are just 5 events with $M \geq 5.0$ and 17, since 1900. That said, the magnitudes of earthquakes of the early 20th century can be overstated. For that reason, two different magnitude thresholds of the earthquake locations being recognized $M_0 = 4.5$ and $M_0 = 5.0$ were considered in the region.

As can be seen from Figure 9d, FCAZ zones are well aligned with the location of the epicenters of significant earthquakes (1900–2008) with $M \geq 4.5$. Only 5 (11.4%) out of 44 epicenters are situated outside of recognized zones. It should be noted that all missed earthquakes occurred before the commencement of the used instrumental catalog. That said, 3 earthquakes have magnitudes $M = 4.5$–4.7 identified to a precision of ± 0.5. For the threshold $M_0 = 4.5$, it is fairly safe to say that there are only two missed targets. FCAZ zones contain 67% of the earthquake epicenters, with $M \geq 3.5$ among those present in the catalog.

The recognized FCAZ zones (Figure 9d) turn out to be connected with the events of higher magnitude threshold $M_0 = 5.0$. At the moment of recognition, there are 17 known earthquakes with $M \geq 5.0$ in the region. The epicenters of 15 (88.2%) of them are located within or at the boundaries of FCAZ zones. Accordingly, if we consider earthquakes with $M \geq 5.0$, then FCAZ zones can be interpreted as areas prone to the same events. That said, two missed targets are the same two Black Sea earthquakes as in the previous reasoning about the threshold $M_0 = 4.5$.

Nine earthquakes with $M \geq 4.5$ occurred in the considered region after the end of the used instrumental catalog. The epicenters of eight of them lie strictly within FCAZ zones. The only missed target error is the epicenter of the earthquake, with $M = 4.6$ located offshore in the Black Sea.

The considered Crimea–Caucasus region is the first one for which FCAZ zones were interpreted for two different magnitude thresholds. In other words, in the recognition problem of the areas prone to earthquakes, there was variation in the magnitude threshold M_0 [60].

The above statistical data allows, to a great extent of reliability, interpreting FCAZ zones (Figure 9) as the areas prone to strong earthquakes in California and the areas prone to significant earthquakes in the Altai–Sayan region, the Caucasus, as well as in the Crimean Peninsula and the northwestern Caucasus.

The description of the first-ever successive recognition of the areas prone to earthquakes for several magnitude thresholds in the same region is given further. This recognition was performed using the SFCAZ method mentioned above, which further develops FCAZ. Successively studied were the areas prone to earthquakes with $M \geq 5.5$, $M \geq 5.75$, and $M \geq 6.0$ in the Baikal–Transbaikal region [35].

Phase one of the research entailed the solution of a classical problem of recognizing the areas prone to significant earthquakes ($M \geq M_0 = 5.5$). Figure 10a shows the recognized zones that are well aligned with the earthquake epicenters with $M \geq 5.5$. Out of 71 such earthquakes, the epicenters of two (2.8%) are located outside of recognized zones, thus creating missed target errors. These two earthquakes occurred before the commencement of active instrumental observations in the region (1929 and 1957) and have a magnitude $M = 5.6$, identified to a precision of ± 0.5 [52], and their epicenters are located outside of the Russian Federation. Accordingly, their actual magnitude can be lower than the threshold $M_0 = 5.5$ and the completed recognition is likely to have no missed targets.

Totally new are the second and third phases of successive recognition. Phase two entailed studying the areas prone to significant earthquakes with $M \geq 5.75$ in the same Baikal–Transbaikal region. To that end, only the epicenters that were included in DPS clusters during phase one were used as recognition objects. Accordingly, inside the DPS clusters that define high seismicity zones for $M \geq 5.5$, subclusters and morphogenetic areas prone to stronger significant earthquakes were recognized.

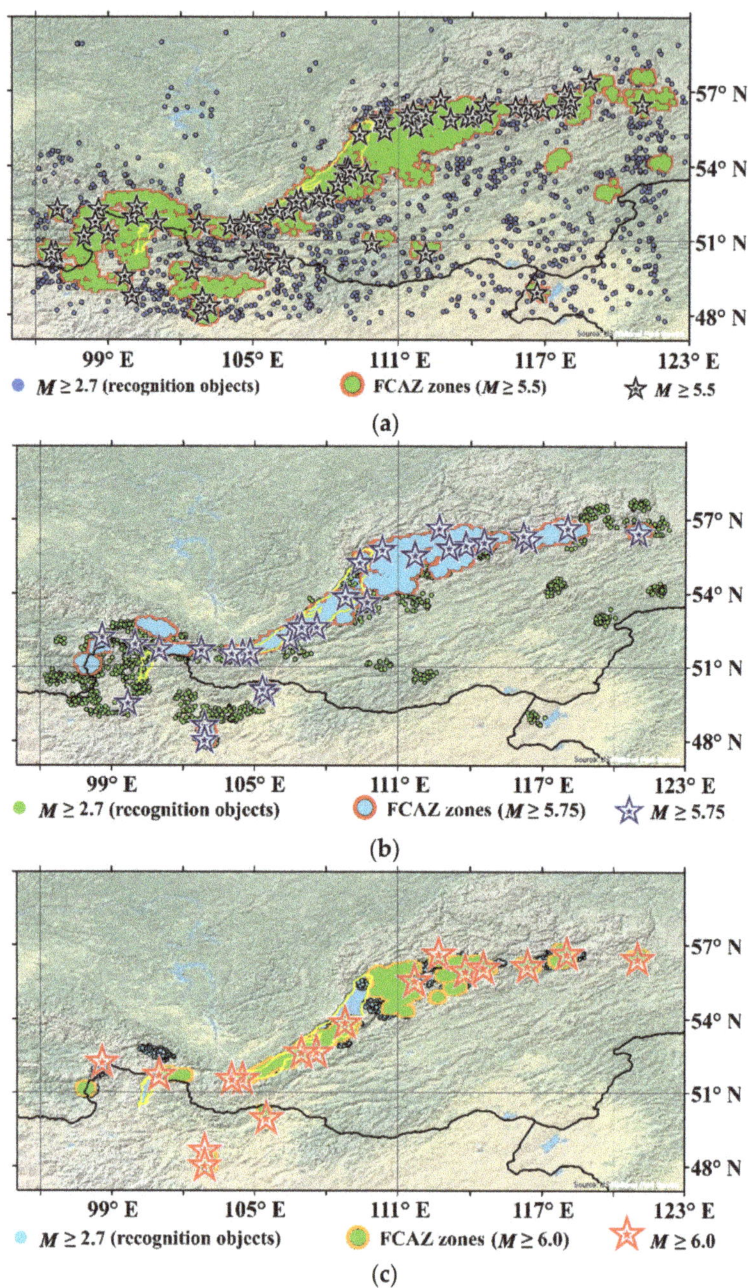

Figure 10. Baikal–Transbaikal region: (**a**) zones prone to earthquakes with $M \geq 5.5$ and the epicenters of earthquakes with $M \geq 5.5$; (**b**) zones prone to earthquakes with $M \geq 5.75$ and the epicenters of earthquakes with $M \geq 5.75$; (**c**) zones prone to earthquakes with $M \geq 6.0$ and the epicenters of earthquakes with $M \geq 6.0$.

The zones recognized in this way are well aligned with the earthquake epicenters with $M \geq 5.75$ (Figure 10b). The epicenters of just 3 (10%) of 30 such significant earthquakes lie outside of their boundaries. These are the epicenters of earthquakes with a fairly inaccurate identification of magnitude: $M = 5.8 \pm 0.5$ and $M = 5.8 \pm 0.2$ [52]. The magnitude of the third earthquake was recalculated from the energy class. The magnitudes of these three earthquakes are highly likely to have the values of $M < 5.75$, and the earthquakes themselves are highly unlikely to constitute the subject matter of research. It should be noted that the recognized territories form part of high seismicity zones for the magnitude threshold $M_0 = 5.5$, identified during phase one of the research.

Phase three entailed recognizing the areas prone to strong earthquakes with $M \geq 6.0$. The epicenters of earthquakes included in the DPS clusters during phase two have already been used as recognition objects. Out of 17 earthquakes with $M \geq 6.0$, the epicenters of just two (11.7%) are located outside of recognized zones (Figure 10c). The first one is the epicenter of the 1939 earthquake with $M = 6.0 \pm 0.3$ [52]; the second one which occurred in 2008, with $M = 6.3$, is located at the distance of $0.15°$ of the mapped zones.

After 2010, 3 earthquakes with $M \geq 5.5$ occurred in the considered region. The epicenters of two of them lie strictly within the zones corresponding to their magnitudes, which is an argument in favor of the reliability of SFCAZ recognition results.

Successive recognition using the SFCAZ method made it possible to obtain a chain of high seismicity areas, in which the zones for greater threshold magnitudes are inserted in the relevant zones for smaller ones. Accordingly, the results of successive recognition can be used in practical seismic zoning. The results of completed successive recognition allow us to argue that the performed transition from FCAZ to SFCAZ does not impair the quality of obtained results [35].

After the end of the used instrumental catalogs, 22 earthquakes with $M \geq M_0$ occurred in 5 considered regions. These events allowed conducting a pure experiment. It should be noted that 19 epicenters (86.3%) are located within high seismicity zones. Such a result of a pure experiment should be recognized as successful. This yielded an objective argument in favor of result reliability for completed FCAZ recognition.

The earthquake with $M = 7.1$, which occurred in California on 6 July 2019 (white star in Figure 9a), deserves a separate mention. The epicenter of this earthquake is located inside FCAZ zones in the territory with no prior strong earthquakes. It should be noted that this epicenter is located outside of the zones recognized by the EPA method [62].

4. Discussion

4.1. Justification of Reliability of FCAZ Recognition Results

Simultaneously with the pure experiment (see above) or in the absence of the same, reliability was assessed based on the computational control experiments. FCAZ recognition employs two types of control experiments—individual seismic history and complete seismic history.

In the individual seismic history experiment, FCAZ zones are constructed based on findings from the DPS clustering of the earthquake epicenters (with $M \geq M_R$) only for 20 years preceding the events with $M \geq M_0$. The experiment ends with an analysis of the relative position of the recognized zones and the epicenter of the earthquake with $M \geq M_0$, for which the zones were constructed.

The complete seismic history experiment excludes from the used instrumental catalog the epicenters for the past few years during which events with $M \geq M_0$ have occurred. FCAZ zones are recognized through the use of DPS clustering of the epicenters remaining in the catalog. The experiment ends with an analysis of the location of the earthquake epicenters with $M \geq M_0$ from the discarded part of the catalog relative to the recognized FCAZ zones.

It should be noted that to improve the objectivity of computational experiments, they are conducted using the same values of the FCAZ method parameters (q, β, δ, ω, v, C) (i.e., the DPS and E^2XT algorithms) as for the main recognition variant. The values β, ω, and v in the main recognition variant (see above) were computed in an automated manner by the artificial intelligence blocks [21].

Series of control experiments were conducted for the mountain belt of the South American Andes, the Pacific Coast of the Kamchatka Peninsula, California, and the Caucasus. Figure 11 shows typical results of experiments in California as an example.

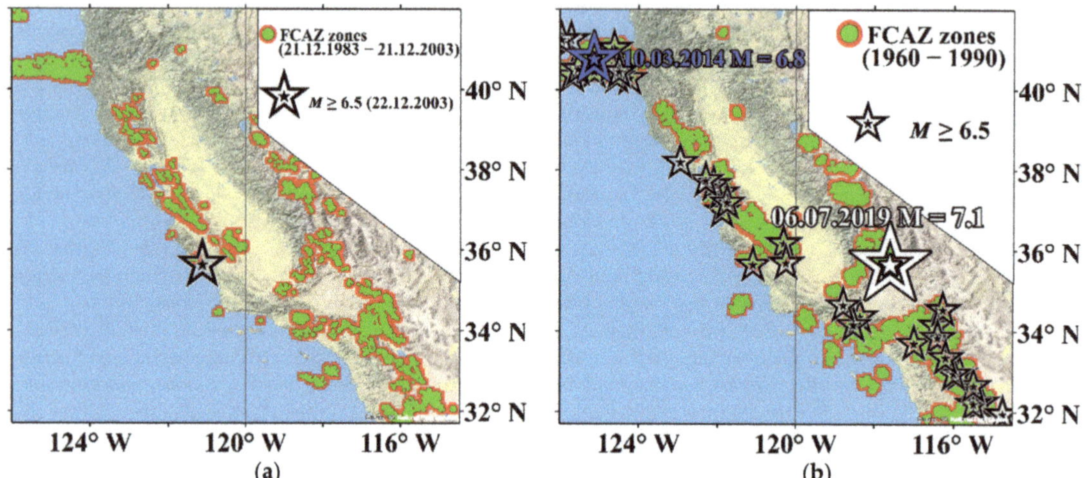

Figure 11. California: (**a**) computational individual seismic history experiment for the earthquake dated 22 December 2003; (**b**) computational complete seismic history experiment (1960–1990) and the epicenters of earthquakes with $M \geq 6.5$.

A comparative analysis of the spatial location of 29 FCAZ zones recognized in the individual seismic history experiments and the main results of FCAZ recognitions (see above) demonstrated a high degree of their similarity. That said, the epicenters of 27 out of 29 earthquakes involved in the experiments are located inside or at the boundaries of the recognized zones.

The FCAZ zones recognized in the course of complete seismic history experiments in terms of their forms and spatial location are close to the FCAZ zones of the main recognition variants. The zones include 25 out of 27 epicenters of earthquakes with $M \geq M_0$, which occurred years later (in particular, 10–25 years) after the date of the last recognition object (earthquake epicenter). For instance, in California, the epicenter of the earthquake with $M = 7.1$ (white star in Figure 11b), which occurred 28.5 years after the end of the catalog used in the experiment is located strictly inside FCAZ zones.

The results of control experiments demonstrate the stability of FCAZ recognition in time and space. This confirms the reliability of the main recognition variants in the studied regions as the zones prone to the strongest, strong, and significant earthquakes.

A comparative analysis of FCAZ zones and the EPA zones recognized earlier [4,62–64] was conducted in the mountain belt of the Andes, on the Pacific Coast of Kamchatka, in California, and the Caucasus. FCAZ zones typically occupy a smaller area than EPA zones. An exception is the mountain belt of the Andes, where FCAZ recognition covered a larger area. Figure 12 shows the comparison of FCAZ zones and EPA zones in Kamchatka and California.

In Kamchatka, high seismicity territories identified by both methods have a common, northeastern strike due to the subduction zone (Figure 12a). That said, FCAZ zones are typically located northwest of EPA zones. This is because most objects recognized as high seismicity ones using the EPA method were formed by the intersection of a deep-water trench with the morphostructural lineaments of rank II and III. At the same time, the main part of the epicenters that represent FCAZ recognition objects are located in the Benioff zone (seismic focal zone) within the continental slope before the trench and are generated by the convergent interaction of two lithospheric plates.

The FCAZ zones are well aligned with the epicenters of known strongest (the Andes and Kamchatka), strong (California), and significant (the Caucasus) earthquakes. A check of this kind of alignment for EPA zones is not a clearly formulated objective. The reason is the construction peculiarities of morphostructural zoning scheme and the selection of EPA recognition objects, especially in the Pacific Seismic Rim regions. Moreover, EPA has no formalized transition from the classification of point objects to sought flat high seismicity zones with unambiguous boundaries. EPA solves this nontrivial problem by the trivial construction of circles with a radius proportionate to the magnitude of recognized earthquakes around the objects classified as high seismicity ones. Circles coincide with the areas initially used to compute the values of characteristics of the objects. The reasonableness of such transition is not obvious.

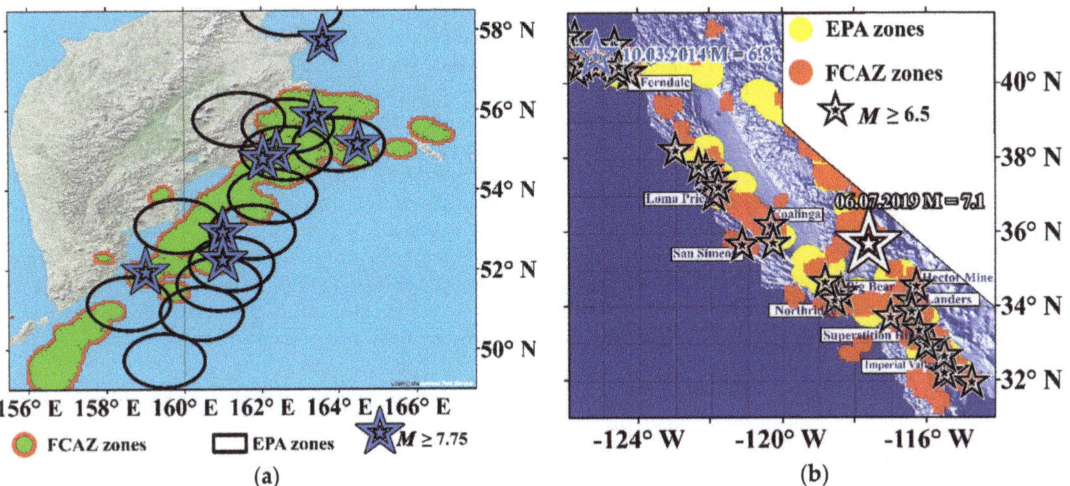

Figure 12. Comparison of the zones prone to earthquakes recognized by FCAZ and EPA methods: (**a**) Pacific Coast of the Kamchatka Peninsula; (**b**) California.

As regards the events with $M \geq M_0$, which constitute the material for pure experiment for both methods (FCAZ and EPA), seven out of eight epicenters of such earthquakes are located inside or at the boundaries of FCAZ zones. That said, only four epicenters are guaranteed to be located inside the EPA zones. It should be noted that the epicenter of the earthquake dated 6 July 2019, with $M = 7.1$ in California is situated strictly inside the FCAZ zones, yet outside of the EPA zones. Summing it up, it is safe to say that the result of FCAZ recognition offers a whole range of benefits as compared with EPA results.

To ascertain the contribution of foreshock and aftershock sequences to the formation of the final result of FCAZ recognition, for the first time, epicenters from declustered catalogs were used as recognition objects. On the Pacific Coast of the Kamchatka Peninsula and in California, the FCAZ zones recognized based on complete and declustered catalogs turned out to be almost coinciding. This evidences that for the considered regions the existence of foreshock and aftershock sequences in the catalogs does not have a significant impact on the results of recognition of high seismicity areas as part of the FCAZ clustering method.

The optimal values of the parameter β computed automatically (the maximality of density in the DPS clusters, and in fact, the algorithm's "look" at the topology of the set of recognition objects and the separability of their dense condensations from the loose complement) for both recognitions in Kamchatka turned out to be very close: -0.2 and -0.2 for the declustered catalog; -0.15 and -0.2 for the complete catalog. Similar optimal values β in California are different. This can be explained by the fact that after a declustering of the catalog, the number of recognition objects went down by 68%, causing a change in the quantitative-spatial distribution of the set of objects. At the same time, the experiment in California can also be treated as successful since the results show that declustering a set of FCAZ recognition objects has not led to a significant change in either the DPS clusters or, in fact, the FCAZ zones [65].

4.2. FCAZ Recognition as the Problem of Advanced Systems Analysis

The FCAZ recognition problem is considered from the standpoint of advanced systems analysis [e.g., https://siiasa.ac.at/ access date: 30 July 2021]. The process and result of identification of potentially high seismicity hazard zones represent a complicated system [66]. The condition of the system depends on both spatial coordinates of recognition objects and on time. The results of FCAZ recognition obtained above follow from the algorithmic analysis of the currently identified objects $W = \{w\}$, which represent numerous epicenters of, generally speaking, fairly weak earthquakes.

For today, FCAZ performed a reliable recognition of sought high seismicity areas in several mountainous countries. Substantiations of such reliability are given for a certain period. This period is not long enough in both geological and real-time. In practice, it means tens, maximum hundreds of years. This period is characterized by the fact that a set of objects $w \in W$ does not change drastically throughout the period. Here, a drastic change means not only the emergence of the clouds of new epicenters of earthquakes with $M \geq M_R$ in previously aseismic areas but also significant alteration of the object distribution topology.

Let Δt be a time interval during that the set W did not undergo any drastic changes. It is natural to assume that the FCAZ result obtained at the moment t_1 will take place until the moment $t_2 = t_1 + \Delta t$. Since t_2, the set W, has significantly changed its spatial form and/or topology. Consequently, at the moment t_2 it is necessary to perform a new FCAZ recognition taking into account the newly received initial data.

Treating this reasoning as the first step of the induction process also makes it easy to determine $\Delta_i t$ and the succession of the pairs:

$$\{(t_i, \text{FCAZ}(t_i)) : i = 1, 2, \ldots\}, \tag{18}$$

where time values t_i are the moments when FCAZ recognitions are repeated.

It should be noted that generally speaking, $\Delta_i t \neq \Delta_j t$, $\forall\, i,\, j = 1, 2, \ldots,\, i \neq j$ studying the dependence of $\Delta_i t$ on changes to the set W over time represents an independent nontrivial problem of systems analysis, which falls beyond the scope of this paper.

Accordingly, an analytical approach to the recognition of potentially high seismicity areas as a complex system that changes over time, even though stable over fairly long local intervals, was created in the present paper. The approach is based on the dynamic changes of the principal parameters of the system. The latter justifies the attribution of algorithmic succession $T(i) \times \text{FCAZ}$, where $T = \{t_i; i = 1, 2, \ldots\}$ is defined by the formula (18), to systems analysis methods. The general scheme of this method is illustrated in Figure 13, where μ_i is the measure of recognition quality at the moment t_i.

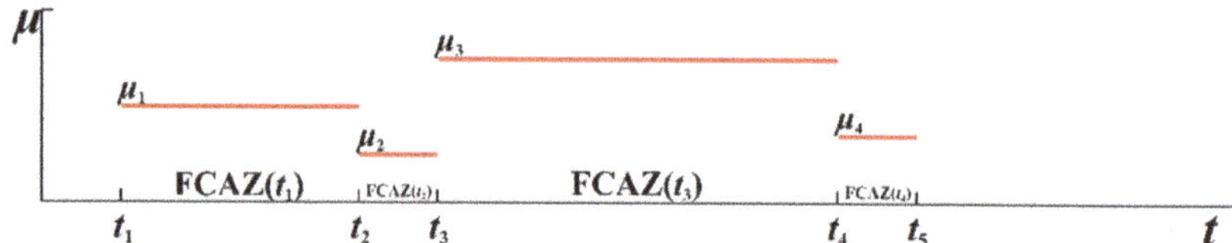

Figure 13. Illustration of the systems analysis method developed based on FCAZ recognition. Deep red color shows the values of the FCAZ recognition quality measure at different time intervals.

Let $\text{FCAZ}_{\gamma_1}(W_{t_1}) : G = B_{t_1} \coprod H_{t_1}$ be the result of FCAZ recognition ($M \geq M_0$) at the moment t_1. That said:

- $W_{t_1} = \{w\}$ is the finite set of recognition objects (the epicenters of earthquakes with $M \geq M_R$) at the moment t_1, $|W_{t_1}| = n_{t_1}$;
- $G = \{g\}$ denotes certain coverage of the considered region by square objects, on which the E^2XT algorithm works;
- $\gamma_1 = \{\delta, C, \omega, v, q, \beta\}$ is a set of values of free parameters of FCAZ selected for an optimal recognition at the moment t_1;
- $B_{t_1} \subset G$ and $H_{t_1} \subset G$ are the subsets of objects $g \in G$ classified as high seismicity and low seismicity ones, respectively, i.e., the objects $g \in B_{t_1}$ are fairly close, and the objects $g \in H_{t_1}$ are fairly distant from known and potential areas prone to strong earthquakes, $B_{t_1} \cup H_{t_1} = G$, $B_{t_1} \cap H_{t_1} = \varnothing$.

Let B_{0,t_1} denote the set of epicenters of strong earthquakes that had occurred by the moment t_1. It is obvious that the higher the value of the inclusion measure of epicenters B_{0,t_1} in the subset of high seismicity objects B_{t_1}, the better FCAZ recognition at the moment t_1:

$$\mu\left(B_{0,t_1} \subset B_{t_1}\right) = \left|B_{0,t_1} \cap B_{t_1}\right| / \left|B_{0,t_1}\right|. \tag{19}$$

The quality of the FCAZ recognition problem considered above is determined by the fact that the results of future (after the appearance of new objects with time) expansions $FCAZ_\gamma(W_t) : G = B_t \coprod H_t$ tend to the limit characterized by the condition:

$$\lim_{t \to \infty} \mu(B_{0,t} \subset B_t) \to 1. \tag{20}$$

Let us assume by the moment $t_2 = t_1 + \Delta t\, Z_{t_1,t_2}$ more strong earthquakes occurred, $B_{0,t_2} = B_{0,t_1} \cup Z_{t_1,t_2}$, i.e., bringing their total number to $\left|B_{0,t_2}\right| = \left|B_{0,t_1} \cup Z_{t_1,t_2}\right|$. That said, the total number of occurring earthquakes with $M \geq M_R$ among the recognition objects increased by z_{t_1,t_2} to a total of $n_{t_2} = n_{t_1} + z_{t_1,t_2}$. Let us denote this new set of objects $W_{t_2} = \{w\}$. In this new situation, at the moment $t_1 + \Delta t$ we have important additional information in place, which was not available to us at the moment t_1. Accordingly, it is necessary to perform FCAZ recognition this time based on W_{t_2} to obtain the expansion $FCAZ_{\gamma_2}(W_{t_2}) : G = B_{t_2} \coprod H_{t_2}$ (Figure 13).

FCAZ recognition is determined by the selection of free parameters $\gamma = \{\delta, C, \omega, v, q, \beta\}$. The parameters of the E^2XT algorithm, as well as its result, directly depend on the recognized DPS clusters. In turn, β in DPS is the maximality level of density of DPS clusters, which depends on the spatial arrangement of objects. Due to the earthquakes with $M \geq M_R$, which occurred over the time $\Delta t = t_2 - t_1$, the spatial distribution of objects $w \in W_{t_2}$ will differ from the distribution of objects $w \in W_{t_1}$. For this reason, the selection of values $\gamma_2 = \{\delta, C, \omega, v, q, \beta\}$ for $FCAZ_{\gamma_2}(W_{t_2}) : G = B_{t_2} \coprod H_{t_2}$ must be performed by the above-mentioned artificial intelligence blocks. These blocks ensure the selection of optimal values of input parameters accounting the spatial distribution of recognition objects at a given moment in time.

It is clear that the spatial distribution of a set of objects W_{t_2} can be so dramatically different from the spatial distribution of W_{t_1} that B_{t_1} will not be a proper subset B_{t_2}. In other words, the threshold (20) can fail to be achieved. To prevent this kind of situation and create a successive monotonous growing of FCAZ zones as the high seismicity areas recognized at the moment t_2, the integration of the zones B_{t_1} and B_{t_2} should be taken, i.e., $B_{t_2} = B_{t_2} \cup B_{t_1}$.

FCAZ recognition in the subsequent moments in time $t_k = t_{k-1} + \Delta t$, $k = 3, 4, \ldots$ is constructed similarly following the process of induction.

Based on the FCAZ results presented in this paper, the moment in time when control experiments were conducted (e.g., complete seismic history), for any of the studied regions is fixed as t_1. Then t_2 is the moment for which the main result of FCAZ recognition was obtained. Then the comparison of sets of recognition objects and FCAZ zones at the moments t_1 and t_2 allows concluding that in respect of all regions considered over the time intervals $\Delta t = t_2 - t_1$, the sets of objects $w \in W$ did not undergo any drastic change. Accordingly, given such fixed t_1 and t_2, the moment t_2 is not yet time for the performance of new FCAZ recognition taking into account new initial data. In this situation, pure examination and computational control experiments gain special importance.

Similarly, the time has not yet come for new FCAZ recognition either if we take t_1 as the moments for which the main results of FCAZ recognition are obtained and take as t_2, for instance, the year 2021.

5. Conclusions

The problem of recognition of the areas prone to strong (with $M \geq M0$) earthquakes [4,5,67,68] is studied in this paper using two methods developed by the authors. Their fundamental difference lies in the selection of recognition objects.

In the first method, objects are vicinities of intersections of lineament axes constructed using a formalized technique of morphostructural zoning. In the second method, objects are constituted by the epicenters of all earthquakes that meet the condition $M \geq MR$, where threshold MR is significantly lower than the magnitude threshold M0 of the recognized earthquake areas.

The methods also differ in the sets of characteristics of object description and the employed pattern recognition algorithms. In the first case, these are geological-geophysical and geomorphologic characteristics and the original Barrier-3 algorithm. In the second case, these are the characteristics of epicenters of weak earthquakes and systems analysis procedure for the objective recognition of dense condensations of FCAZ.

Despite the critical differences between these two original methods, their recognition results are well aligned in the Altai–Sayan–Baikal region and the Caucasus. The territories classified as high seismicity ones by both methods should be viewed as the most hazardous since they are recognized as such by independent methods based on different recognition objects and their characteristics.

The first method allows, from the standpoint of dynamic systems analysis, repeatedly solving the problem of classification of lineament intersections into high and low seismicity ones. This relies on the fact that learning is every time performed only for one high seismicity class, which is easy to form with due regard for new strong earthquakes that have occurred. This, in turn, contributed significantly to the development of the classical EPA approach towards the recognition of high seismicity areas [3–5,67,69,70].

Previously, there was a problem of identification of the learning set of the objects in whose vicinities strong earthquakes cannot occur. This problem is solved in the paper by developing an original method for image recognition called Barrier-3.

This algorithm makes it possible to classify objects into high and low seismicity based on one learning class. Barrier-3, having information about the objects with known epicenters of earthquakes with $M \geq M0$ in their vicinities, enables finding a set of the so-called similar objects.

The recognition of the areas prone to earthquakes is the first developed method to rely on the hypothesis about the association of epicenters of strong earthquakes with the intersections of morphostructural lineaments, which was confirmed in [18]. Accordingly, building the morphostructural zoning map is an important phase of the first method for studying the problem. That said, despite the logical formalization conducted as early as 1977 by a group of mathematicians under the guidance of I.M. Gelfand, the process of morphostructural zoning remains ambiguous. In this regard, a question was pending: Can the recognition of strong earthquake-prone areas be performed without constructing morphostructural zoning model? [22]. This paper answers this question positively based on the use of the systems analysis method FCAZ.

The employment of DMA algorithms in this paper, which use the epicenters of earthquakes as recognition objects, justifies this positive answer. Accordingly, the system FCAZ approach is a new step in the study of a recognition problem of strong earthquake-prone areas.

The recognition process of the high seismicity hazard zones in tectonically active regions represents a complicated system. The condition of the system depends on both spatial coordinates of recognition objects and on time. In this regard, FCAZ recognition is viewed in this paper from the perspective of the systems analysis. The system–mathematical model of FCAZ recognition as a complicated dynamic system was developed. The space-and-time model T(i) × FCAZ for recognition of the areas prone to the strongest, strong, and significant earthquakes makes it possible to develop a schedule of subsequent iterations for the recognition of high seismicity hazard zones for the regions studied in this paper.

The following regions with varying seismicity levels were studied by Barrier-3 and FCAZ methods in this paper:

- Barrier-3—the Altai–Sayan–Baikal region ($M \geq 6.0$) and the Caucasus ($M \geq 6.0$).
- FCAZ—mountain belt of the South American Andes ($M \geq 7.75$), the Pacific Coast of the Kamchatka Peninsula ($M \geq 7.75$), and the Kuril Islands ($M \geq 7.75$); California ($M \geq 6.5$); the Baikal–Transbaikal region ($M \geq 5.5, M \geq 5.75, M \geq 6.0$); the Altai–Sayan region ($M \geq 5.5$); the Caucasus ($M \geq 5.0$); the Crimean Peninsula and northwestern Caucasus ($M \geq 4.5, M \geq 5.0$).

The Altai–Sayan–Baikal region, the Pacific Coast of the Kuril Islands, and the Crimean Peninsula were first studied with the employment of methods for the recognition of earthquake-prone areas. Moreover, the Baikal–Transbaikal region was used as an example of the first recognition of earthquake-prone areas for the finite succession of growing magnitude thresholds $M_0^1 < M_0^2 < M_0^3$. The joint presentation of the recognition results obtained by the Barrier-3 and Cora-3 algorithms in the Caucasus based on their composition with a fuzzy set allowed halving the number of missed targets.

It was shown, using California and the Pacific Coast of the Kamchatka Peninsula as an example, that the existence of foreshock and aftershock sequences in the catalogs of earthquakes does not have a significant impact on the FCAZ recognition results. A totality of control experiments conducted in this paper demonstrates the reliability and reproducibility of the interpretation of FCAZ zones as the areas prone to the strongest, strong, and significant earthquakes.

In the studied regions, FCAZ zones occupy a relatively small area as compared with the total seismicity field, which makes up 30%–40% of the total seismicity space and 50%–65% of the space where earthquakes with $M \geq M_R$ occur. This illustrates the spatial nontriviality of the obtained results.

Findings from the paper also demonstrate that low seismicity can actually "manifest" the properties of geophysical fields, which in the classical EPA approach are used directly as the characteristics of recognition objects.

Author Contributions: Conceptualization, B.A.D. and A.D.G.; data curation, B.A.D.; formal analysis, B.A.D., A.D.G., and S.M.A.; funding acquisition, B.A.D.; investigation, B.A.D. and A.D.G.; methodology, B.A.D., A.D.G., and S.M.A.; project administration, B.A.D.; resources, J.K.K. and B.V.D.; software, B.A.D. and I.O.B.; validation, B.A.D., A.D.G., J.K.K., and B.V.D.; visualization, B.A.D. and Y.V.B.; writing—original draft, B.A.D., A.D.G., B.V.D., and Y.V.B. All authors have read and agreed to the published version of the manuscript.

Funding: The reported study was funded by RFBR, project number 20-35-70054 «Systems approach to recognition algorithms for seismic hazard assessment».

Acknowledgments: This work employed data provided by the Shared Research Facility «Analytical Geomagnetic Data Center» of the Geophysical Center of RAS (http://ckp.gcras.ru/ access date: 30 July 2021).

Conflicts of Interest: The authors declare no conflict of interest.

Abbreviations

The following abbreviations are used in this manuscript:

ANSS	Advanced National Seismic System
Barrier-3	Pattern recognition algorithm with one learning class
Cora-3	Pattern recognition algorithm with two learning classes (the most common dichotomy algorithm in the EPA approach)
DMA	Discrete Mathematical Analysis
DPS	Discrete Perfect Sets (algorithm in the structure of the FCAZ method)
E^2XT	Extension (algorithm in the structure of the FCAZ method)
EPA	Earthquake-Prone Areas
FCAZ	Formalized Clustering and Zoning
MSZ	Morphostructural zoning
SFCAZ	Successive Formalized Clustering and Zoning
Top 3	Rank of three strongest characteristics in Barrier-3 algorithm
M	Magnitude
M_0	Magnitude threshold of strong earthquakes
M_R	Magnitude threshold, starting from which the epicenters were used as recognition objects in FCAZ method

References

1. Gelfand, I.M.; Guberman, S.; Izvekova, M.L.; Keilis-Borok, V.I.; Ranzman, E.I. Criteria of high seismicity determined by pattern recognition. *Tectonophysics* **1972**, *13*, 415–422. [CrossRef]
2. Gelfand, I.M.; Guberman, S.A.; Izvekova, M.L.; Keilis-Borok, V.I.; Rantzman, E.Y. On the criteria of high seismicity. *Dokl. Akad. Nauk SSSR* **1972**, *202*, 1317–1320. (In Russian)
3. Gorshkov, A.; Novikova, O. Estimating the validity of the recognition results of earthquake-prone areas using the ArcMap. *Acta Geophys.* **2018**, *66*, 843–853. [CrossRef]
4. Gvishiani, A.D.; Gorshkov, A.I.; Rantsman, E.Y.; Cisternas, A.; Soloviev, A.A. *In Recognition of Earthquake-Prone Areas in the Regions of Moderate Seismicity*; Nauka: Moscow, Russia, 1988; 176p. (In Russian)
5. Gvishiani, A.D.; Soloviev, A.A.; Dzeboev, B.A. Problem of Recognition of Strong-Earthquake-Prone Areas: A State-of-the-Art Review. *Izv. Phys. Solid Earth* **2020**, *56*, 1–23. [CrossRef]
6. Kossobokov, V.G.; Soloviev, A.A. Pattern recognition in problems of seismic hazard assessment. *Chebyshevskii Sb.* **2018**, *19*, 55–90. (In Russian) [CrossRef]
7. Soloviev, A.A.; Gvishiani, A.D.; Gorshkov, A.I.; Dobrovolsky, M.N.; Novikova, O.V. Recognition of earthquake-prone areas: Methodology and analysis of the results. *Izv. Phys. Solid Earth* **2014**, *50*, 151–168. [CrossRef]
8. Alekseevskaya, M.; Gabrielov, A.; Gelfand, I.; Gvishiani, A.; Rantsman, E. Formal morphostructural zoning of mountain territories. *Geophysics* **1977**, *42*, 227–233.
9. Gabrielov, A.M.; Gorshkov, V.I.; Rantsman, E.Y. The experience of morphostructural zoning based on formalized features. In *Recognition and Spectral Analysis in Seismology, 10 of Computational Seismology*; Keilis-Borok, V.I., Ed.; Nauka: Moscow, Russia, 1977; pp. 50–58. (In Russian)
10. Gorshkov, A.I.; Soloviev, A.A.; Zharkikh, Y.I. A Morphostructural Zoning of the Mountainous Crimea and the Possible Locations of Future Earthquakes. *J. Volcanol. Seismol.* **2017**, *11*, 407–412. [CrossRef]
11. Rantsman, E.Y. *Locations of Earthquakes and Morphological Structure of Mountain Countries*; Nauka: Moscow, Russia, 1979; 172p. (In Russian)
12. Rantsman, E.Y.; Glasko, M.P. *Morphostructural Nodes are Locations of Extreme Natural Phenomena*; Media-Press: Moscow, Russia, 2004; 224p. (In Russian)
13. Gvishiani, A.D.; Agayan, S.M.; Dzeboev, B.A.; Belov, I.O. Recognition of Strong Earthquake–Prone Areas with a Single Learning Class. *Dokl. Earth Sci.* **2017**, *474*, 546–551. [CrossRef]
14. Gvishiani, A.D.; Gurvich, V.A. *Dynamical Problems of Classification and Convex Programming: Applications*; Nauka: Moscow, Russia, 1992; 360p. (In Russian)
15. Gvishiani, A.D.; Gurvich, V.A. Time stability of a prediction of sites of strong earthquakes: II. The eastern part of Central Asia. *Izv. Phys. Solid Earth* **1983**, *18*, 665–671.
16. Dzeboev, B.A.; Gvishiani, A.D.; Belov, I.O.; Agayan, S.M.; Tatarinov, V.N.; Barykina, Y.V. Strong Earthquake-Prone Areas Recognition Based on an Algorithm with a Single Pure Training Class: I. Altai-Sayan-Baikal Region, $M \geq 6.0$. *Izv. Phys. Solid Earth* **2019**, *55*, 563–575. [CrossRef]
17. Dzeboev, B.A.; Soloviev, A.A.; Dzeranov, B.V.; Karapetyan, J.K.; Sergeeva, N.A. Strong earthquake-prone areas recognition based on the algorithm with a single pure training class. II. Caucasus, $M \geq 6.0$. Variable EPA method. *Russ. J. Earth Sci.* **2019**, *19*, ES6005. [CrossRef]

18. Gvishiani, A.D.; Soloviev, A.A. On the concentration of major earthquakes round the interactions of morphostructural lineaments in South America. In *Computational Seismology, 13: Interpretation of Seismological Data: Methods and Algorithms*; Keilis-Borok, V.I., Levshin, A.L., Eds.; Allerton: New York, NY, USA, 1981; pp. 42–48.

19. Soloviev, A.A.; Krasnoperov, R.I.; Nikolov, B.P.; Zharkikh, J.I.; Agayan, S.M. Web-Oriented Software System for Analysis of Spatial Geophysical Data Using Geoinformatics Methods. *Izv. Atmos. Ocean Phys.* **2018**, *54*, 1312–1319. [CrossRef]

20. Soloviev, A.A.; Gvishiani, A.D.; Nikolov, B.P.; Nikolova, Y.I. GIS-Oriented Database on Seismic Hazard Assessment for Caucasian and Crimean Regions. *Izv. Atmos. Ocean Phys.* **2018**, *54*, 1363–1373. [CrossRef]

21. Gvishiani, A.D.; Dzeboev, B.A.; Agayan, S.M. FCAZm intelligent recognition system for locating areas prone to strong earthquakes in the Andean and Caucasian mountain belts. *Izv. Phys. Solid Earth* **2016**, *52*, 461–491. [CrossRef]

22. Gvishiani, A.; Dzeboev, B.; Agayan, S. A new approach to recognition of the earthquake-prone areas in the Caucasus. *Izv. Phys. Solid Earth* **2013**, *49*, 747–766. [CrossRef]

23. Agayan, S.M.; Bogoutdinov, S.R.; Gvishiani, A.D.; Kagan, A.I. Smoothing of time series by the methods of discrete mathematical analysis. *Russ. J. Earth Sci.* **2010**, *11*, RE40001. [CrossRef]

24. Agayan, S.M.; Bogoutdinov, S.R.; Krasnoperov, R.I. Short introduction into DMA. *Russ. J. Earth Sci.* **2018**, *18*, ES2001. [CrossRef]

25. Agayan, S.M.; Tatarinov, V.N.; Gvishiani, A.D.; Bogoutdinov, S.R.; Belov, I.O. FDPS algorithm in stability assessment of the Earth's crust structural tectonic blocks. *Russ. J. Earth Sci.* **2020**, *20*, ES6014. [CrossRef]

26. Bogoutdinov, S.R.; Agayan, S.M.; Gvishiani, A.D.; Graeva, E.M.; Rodkin, M.V.; Zlotnicki, J.; Le Mouel, J.L. Fuzzy logic algorithms in the analysis of electrotelluric data with reference to monitoring of volcanic activity. *Izv. Phys. Solid Earth* **2007**, *43*, 597–609. [CrossRef]

27. Gvishiani, A.D.; Agayan, S.M.; Bogoutdinov, S.R. Discrete mathematical analysis and monitoring of volcanoes. *Inzhenernaya Ekol.* **2008**, *5*, 26–31. (In Russian)

28. Gvishiani, A.D.; Agayan, S.M.; Bogoutdinov, S.R. Mathematical Methods of Geoinformatics. I. A New Approach to Clusterization. *Cybern. Syst. Anal.* **2002**, *38*, 238–254. [CrossRef]

29. Gvishiani, A.D.; Agayan, S.M.; Bogoutdinov, S.R.; Soloviev, A.A. Discrete mathematical analysis and applications in geology and geophysics. *Vestn. Kamchatskoi Reg. Organ. Uchebno-Nauchnyi Tsentr. Seriya: Nauk. O Zemle* **2010**, *2*, 109–125. (In Russian)

30. Soloviev, A.A.; Agayan, S.M.; Gvishiani, A.D.; Bogoutdinov, S.R.; Chulliat, A. Recognition of disturbances with specified morphology in time series: Part 2. Spikes on 1-s magnetograms. *Izv. Phys. Solid Earth* **2012**, *48*, 395–409. [CrossRef]

31. Agayan, S.; Soloviev, A. Recognition of dense areas in metric spaces basing on crystallization. *Syst. Res. Inf. Technol.* **2004**, *2*, 7–23.

32. Agayan, S.M.; Bogoutdinov, S.R.; Dobrovolsky, M.N. Discrete perfect sets and their application in cluster analysis. *Cybern. Syst. Anal.* **2014**, *50*, 176–190. [CrossRef]

33. Dzeboev, B.A. A New Approach to Monitoring Seismic Activity: California Case Study. *Dokl. Earth Sci.* **2017**, *473*, 338–341. [CrossRef]

34. Dzeboev, B.A.; Krasnoperov, R.I. On the monitoring of seismic activity using the algorithms of discrete mathematical analysis. *Russ. J. Earth Sci.* **2018**, *18*, ES3003. [CrossRef]

35. Gvishiani, A.D.; Dzeboev, B.A.; Belov, I.O.; Sergeyeva, N.A.; Vavilin, E.V. Successive Recognition of Significant and Strong Earthquake-Prone Areas: The Baikal–Transbaikal Region. *Dokl. Earth Sci.* **2017**, *477*, 1488–1493. [CrossRef]

36. Gorshkov, A.I.; Soloviev, A.A. Recognition of earthquake-prone areas in the Altai-Sayan-Baikal region based on the morphostructural zoning. *Russ. J. Earth Sci.* **2021**, *21*, ES1005. [CrossRef]

37. Gorshkov, A.I.; Soloviev, A.A.; Zharkikh, J.I. Recognition of Strong Earthquake Prone Areas in the Altai–Sayan–Baikal Region. *Dokl. Earth Sci.* **2018**, *479*, 412–414. [CrossRef]

38. Bongard, M.M. *Recognition Problem*; Nauka: Moscow, Russia, 1967; 320p. (In Russian)

39. Bongard, M.M.; Vaintsvaig, M.N.; Guberman, S.A.; Izvekova, M.L.; Smirnov, M.S. Using a learning program for identifying oil reservoirs. *Geol. I Geofiz.* **1966**, *2*, 15–29. (In Russian)

40. Soloviev, A.A.; Novikova, O.V.; Gorshkov, A.I.; Piotrovskaya, E.P. Recognition of potential sources of strong earthquakes in the Caucasus region using GIS technologies. *Dokl. Earth Sci.* **2013**, *450*, 658–660. [CrossRef]

41. Soloviev, A.A.; Gorshkov, A.I.; Soloviev, A.A. Application of the data on the lithospheric magnetic anomalies in the problem of recognizing the earthquake prone areas. *Izv. Phys. Solid Earth* **2016**, *52*, 803–809. [CrossRef]

42. Shebalin, N.V.; Tatevosian, R.E. Catalogue of large historical earthquakes of the Caucasus. In *Historical and prehistorical earthquakes in the Caucasus*; Giordini, D., Balassanian, S., Eds.; NATO ASI Series, 2. Enviroment—Vol. 28; Kluwer Academic Publishers: Dordrecht, The Netherlands; Boston, MA, USA; London, UK, 1997; pp. 201–232.

43. Karapetyan, J.K.; Sargsyan, R.S.; Kazaryan, K.S.; Dzeranov, B.V.; Dzeboev, B.A.; Karapetyan, R.K. Current state of exploration and actual problems of tectonics, seismology and seismotectonics of Armenia. *Russ. J. Earth Sci.* **2020**, *20*, ES2005. [CrossRef]

44. Zadeh, L.A. Fuzzy sets. *Inf. Control* **1965**, *8*, 338–353. [CrossRef]

45. Gvishiani, A.D.; Dzeboev, B.A.; Agayan, S.M.; Belov, I.O.; Nikolova, J.I. Fuzzy Sets of High Seismicity Intersections of Morphostructural Lineaments in the Caucasus and in the Altai–Sayan–Baikal Region. *J. Volcanol. Seismol.* **2021**, *15*, 73–79. [CrossRef]

46. Reasenberg, P. Second-order moment of central California seismicity, 1969−82. *J. Geophys. Res.* **1985**, *90*, 5479–5495. [CrossRef]

47. Zaliapin, I.; Ben-Zion, Y. Earthquake clusters in southern California I: Identification and stability. *J. Geophys. Res. Solid Earth* **2013**, *118*, 2847–2864. [CrossRef]

48. Zaliapin, I.; Gabrielov, A.; Keilis-Borok, V.; Wong, H. Clustering analysis of seismicity and aftershock identification. *Phys. Rev. Lett.* **2008**, *101*, 018501. [CrossRef]

49. Abubakirov, I.R.; Gusev, A.A.; Guseva, E.M.; Pavlov, V.M.; Skorkina, A.A. Mass determination of moment magnitudes M_W and establishing the relationship between M_W and M_L for moderate and small Kamchatka earthquakes. *Izv. Phys. Solid Earth* **2018**, *54*, 33–47. [CrossRef]

50. Dzeboev, B.A.; Agayan, S.M.; Zharkikh, Y.I.; Krasnoperov, R.I.; Barykina, Y.V. Strongest Earthquake-Prone Areas in Kamchatka. *Izv. Phys. Solid Earth* **2018**, *54*, 284–291. [CrossRef]

51. Skorkina, A.A. Scaling of two corner frequencies of source spectra for earthquakes of the Bering fault. *Russ. J. Earth Sci.* **2020**, *20*, ES2001. [CrossRef]

52. Kondorskaya, N.V.; Shebalin, N.V.; Khrometskaya, Y.A.; Gvishiani, A.D. *New Catalog of Strong Earthquakes in the USSR from Ancient Times through 1977*; World Data Center A for Solid Earth Geophysics, Report SE-31; NOAA: Boulder, CO, USA, 1982; 608p.

53. Fedotov, S.A.; Solomatin, A.V. The long-term earthquake forecast for the Kuril-Kamchatka island arc for the September 2013 to August 2018 period; the seismicity of the arc during preceding deep-focus earthquakes in the sea of Okhotsk (in 2008, 2012, and 2013 at $M = 7.7, 7.7$, and 8.3). *J. Volcanol. Seismol.* **2015**, *9*, 65–80. [CrossRef]

54. Dzeboev, B.A.; Krasnoperov, R.I.; Belov, I.O.; Barykina, Y.I.; Vavilin, E.V. Modified algorithmic system FCAZm and strong earthquake-prone areas in California. *Geoinformatika* **2018**, *2*, 2–8. (In Russian)

55. Gvishiani, A.; Dobrovolsky, M.; Agayan, S.; Dzeboev, B. Fuzzy-based clustering of epicenters and strong earthquake-prone areas. *Environ. Eng. Manag. J.* **2013**, *12*, 1–10. [CrossRef]

56. Gvishiani, A.D.; Agayan, S.M.; Dobrovolsky, M.N.; Dzeboev, B.A. Objective epicenter classification and recognition of strong-earthquake-prone areas in California. *Geoinformatika* **2013**, *2*, 44–57. (In Russian)

57. Gvishiani, A.D.; Dzeboev, B.A.; Sergeyeva, N.A.; Belov, I.O.; Rybkina, A.I. Significant Earthquake-Prone Areas in the Altai–Sayan Region. *Izv. Phys. Solid Earth* **2018**, *54*, 406–414. [CrossRef]

58. Gvishiani, A.D.; Dzeboev, B.A. Assessment of seismic hazard in choosing of a radioactive waste disposal location. *Min. J.* **2015**, *10*, 39–43. (In Russian) [CrossRef]

59. Karapetyan, J.K.; Gasparyan, A.S.; Shakhparonyan, S.R.; Karapetyan, R.K. Registration and spectral analysis of waveforms of 10.24.2019 earthquake in the Caucasus using the new IGES-006 seismic sensor. *Russ. J. Earth Sci.* **2020**, *20*, ES6006. [CrossRef]

60. Gvishiani, A.D.; Dzeboev, B.A.; Sergeyeva, N.A.; Rybkina, A.I. Formalized Clustering and the Significant Earthquake-Prone Areas in the Crimean Peninsula and Northwest Caucasus. *Izv. Phys. Solid Earth* **2017**, *53*, 353–365. [CrossRef]

61. Nekrasova, A.K.; Kossobokov, V.G. Unified scaling law for earthquakes in Crimea and Northern Caucasus. *Dokl. Earth Sci.* **2016**, *470*, 1056–1058. [CrossRef]

62. Gelfand, I.M.; Guberman, S.A.; Keilis-Borok, V.I.; Knopoff, L.; Press, F.S.; Ranzman, E.Y.; Rotwain, I.M.; Sadovsky, A.M. Pattern recognition applied to earthquake epicenters in California. *Phys. Earth Planet. Inter.* **1976**, *11*, 227–283. [CrossRef]

63. Gvishiani, A.D.; Zhidkov, M.P.; Soloviev, A.A. On transferring the criteria of high seismicity of Andean mountain belt to Kamchatka. *Izv. Akad. Nauk SSSR. Fiz. Zemli* **1984**, *1*, 20–33. (In Russian)

64. Gvishiani, A.D.; Zhidkov, M.P.; Soloviev, A.A. Recognition of strong-earthquake-prone areas: X. M 7.75 earthquake prone areas on the Pacific Coast of South America. In *Mathematical Models of the Structure of the Earth and the Earthquake Prediction, 14 of Computational Seismology*; Keilis-Borok, V.I., Ed.; Allerton: New York, NY, USA, 1983; pp. 56–68.

65. Dzeboev, B.A.; Karapetyan, J.K.; Aronov, G.A.; Dzeranov, B.V.; Kudin, D.V.; Karapetyan, R.K.; Vavilin, E.V. FCAZ-recognition based on declustered earthquake catalogs. *Russ. J. Earth Sci.* **2020**, *20*, ES6010. [CrossRef]

66. Zgurovsky, M.Z.; Pankratova, N.D. *System Analysis: Theory and Applications*; Data and Knowledge in a Changing World; Springer: Berlin/Heidelberg, Germany, 2007; 447p. [CrossRef]

67. Cisternas, A.; Godefroy, P.; Gvishiani, A.; Gorshkov, A.; Kossobokov, V.; Lambert, M.; Ranzman, E.; Sallantin, J.; Saldano, H.; Soloviev, A.; et al. A dual approach to recognition of earthquake prone areas in the Western Alps. *Ann. Geophys.* **1985**, *3*, 249–270.

68. Gorshkov, A.; Kossobokov, V.; Soloviev, A. Recognition of earthquake-prone areas. In *Nonlinear Dynamics of the Lithosphere and Earthquake Prediction*; Keilis-Borok, V., Soloviev, A., Eds.; Springer: Heidelberg, Germany, 2003; pp. 239–310. [CrossRef]

69. Gorshkov, A.I.; Kuznetsov, I.V.; Soloviev, A.A.; Panza, G.F. Identification of future earthquake sources in the Carpatho-Balkan orogenic belt using morphostuctural criteria. *Pure Appl. Geophys.* **2000**, *157*, 79–95. [CrossRef]

70. Gorshkov, A.I.; Panza, G.F.; Soloviev, A.A.; Aoudia, A. Morphostructural zonation and preliminary recognition of seismogenic nodes around the Adria margin in peninsular Italy and Sicily. *J. Seismol. Earthq. Eng.* **2002**, *4*, 1–24.

applied
sciences

MDPI

Article

The Analysis of the Aftershock Sequences of the Recent Mainshocks in Alaska

Mohammadamin Sedghizadeh * and Robert Shcherbakov

Department of Earth Sciences, University of Western Ontario, London, ON N6A 5B7, Canada; rshcherb@uwo.ca
* Correspondence: msedghiz@uwo.ca

Abstract: The forecasting of the evolution of natural hazards is an important and critical problem in natural sciences and engineering. Earthquake forecasting is one such example and is a difficult task due to the complexity of the occurrence of earthquakes. Since earthquake forecasting is typically based on the seismic history of a given region, the analysis of the past seismicity plays a critical role in modern statistical seismology. In this respect, the recent three significant mainshocks that occurred in Alaska (the 2002, Mw 7.9 Denali; the 2018, Mw 7.9 Kodiak; and the 2018, Mw 7.1 Anchorage earthquakes) presented an opportunity to analyze these sequences in detail. This included the modelling of the frequency-magnitude statistics of the corresponding aftershock sequences. In addition, the aftershock occurrence rates were modelled using the Omori–Utsu (OU) law and the Epidemic Type Aftershock Sequence (ETAS) model. For each sequence, the calculation of the probability to have the largest expected aftershock during a given forecasting time interval was performed using both the extreme value theory and the Bayesian predictive framework. For the Bayesian approach, the Markov Chain Monte Carlo (MCMC) sampling of the posterior distribution was performed to generate the chains of the model parameters. These MCMC chains were used to simulate the models forward in time to compute the predictive distributions. The calculation of the probabilities to have the largest expected aftershock to be above a certain magnitude after a mainshock using the Bayesian predictive framework fully takes into account the uncertainties of the model parameters. Moreover, in order to investigate the credibility of the obtained forecasts, several statistical tests were conducted to compare the performance of the earthquake rate models based on the OU formula and the ETAS model. The results indicate that the Bayesian approach combined with the ETAS model produced more robust results than the standard approach based on the extreme value distribution and the OU law.

Keywords: epidemic type aftershock sequence model; extreme value distribution; Bayesian predictive distribution

Citation: Sedghizadeh, M.; Shcherbakov, R. The Analysis of the Aftershock Sequences of the Recent Mainshocks in Alaska. *Appl. Sci.* **2022**, *12*, 1809. https://doi.org/10.3390/app12041809

Academic Editor: Amadeo Benavent-Climent

Received: 19 December 2021
Accepted: 5 February 2022
Published: 10 February 2022

Publisher's Note: MDPI stays neutral with regard to jurisdictional claims in published maps and institutional affiliations.

1. Introduction

The Pacific Ring of Fire is one of the most seismically active regions of the world. Alaska and western Canada are a part of this ring and are prone to the occurrence of significant earthquakes. This geographic region is characterized by high seismic activity and is capable of producing megathrust earthquakes. These earthquakes can pose significant hazard and are also capable of triggering tsunamis or intense ground shaking [1] and subsidiary hazards such as liquefaction, landslides and aftershocks [2]. While tsunamis pose a serious threat to coastal areas, ground shaking can cause damage to infrastructure and endanger human life. Therefore, it is important to perform a comprehensive statistical analysis of the aftershock sequences in the Aleutian subduction zone and central Alaska. Moreover, the occurrence of large aftershocks poses a significant risk to the infrastructure that has been affected by a mainshock. Therefore, estimating the probabilities for the occurrence of the largest expected aftershocks plays an important role in post-mainshock decision-making [3,4].

One of the earliest empirical studies of the difference between the magnitude of the mainshock and its largest aftershock was conducted by Båth [5], who postulated that the largest aftershock is on average 1.2 magnitudes lower than the mainshock regardless of the magnitude of the mainshock. Vere-Jones [6,7] proposed that the magnitude difference between the mainshock and the largest aftershock was independent of the number of events. Reasenberg and Jones [8] were one of the first in developing an aftershock forecasting model. They introduced a parametric model that was capable of computing the probabilities of aftershocks in a certain time window after a mainshock for California. Michael et al. [9] proposed the methodology which aforementioned model parameters can be estimated with Bayesian updating from both the ongoing aftershock sequence and from historic aftershock sequences.

An important step in the calculation of the probability of having an earthquake above a certain magnitude is the estimation of the model parameters that describe the seismicity rate and the frequency-magnitude distribution. Those parameters are highly dependent on the lower cut-off magnitude, m_0. The correct estimation of the cut-off magnitude plays a crucial role in earthquake forecasting and modelling. Mignan and Woessner [10] emphasized that a high-value of the cut-off magnitude can result in under-sampling of useful data and a low-value of the cut-off magnitude can result in uncertainty and bias of the estimated seismicity parameters and forecasting model, respectively.

The other issue in aftershock forecast modelling is the catalogue incompleteness right after the occurrence of strong mainshocks [11,12]. This early catalogue incompleteness can affect significantly the estimation of the parameters of the earthquake decay rate. The uncertainties in the estimation of the parameters of the aftershock decay rate can result in significant miscalculation of the probabilities for the occurrence of largest events. The empirical prior probability distribution was presented by Omi et al. [13] to reduce the uncertainty of the parameter estimation of the ETAS model regarding the incompleteness of the earthquake catalogues.

Utilizing generic parameters to create an aftershock forecast model for the early days after the mainshocks is one of the possible ways to control the catalogue incompleteness. Page et al. [14] introduced a method for generic parameter estimation by using tectonic zoning of García et al. [15] to improve the spatial distribution of forecasted events. In this approach, Bayes' rule and aftershock records are used to update the generic parameters. In addition, the distribution of the regional generic parameters can be considered as a prior and the aftershock data can be used to calculate the posterior distribution. Michael et al. [9] applied this approach to the 2018 Anchorage aftershock sequence. They reported that the use of the generic parameters for the forecast model leads to the overestimation of the seismic activity.

One of the critical tasks in statistical seismology is the ability to accurately and reliably forecast the evolution of earthquake sequences. A consistent approach for earthquake forecast testing has been implemented in the Collaboratory for the Study of Earthquake Predictability (CSEP) [16–19]. In this framework, the gridded rate forecast is used in which the selected geographic area is separated into zones then the number of earthquakes in each zone is estimated [19]. In addition, the number of earthquakes in each forecast bin is considered to be independent of the other bins and follows the Poisson distribution. Several statistical tests were developed as part of the CSEP framework to examine earthquake forecasts. As a result of these developments, it is possible to determine if a particular forecasting scheme is able to accurately replicate locations, magnitudes, and the observed numbers of earthquakes [19,20]. Various forecasting algorithms can also be compared using the aforementioned likelihood-based tests. For example, retrospective aftershock forecasting of the 2011 Tohoku, Japan; 2010 Canterbury, New Zealand; 2016 Kaikoura, New Zealand; and 2019 Ridgecrest, California earthquakes were tested by using this approach [4,21–24].

In this study, the analysis of three major earthquake sequences that occurred in Alaska in the past 20 years was conducted to test retrospectively the ability to forecast the magnitudes of the largest expected aftershocks. Specifically, the 23 January 2018 Mw 7.9 Kodiak earthquake occurred in the Gulf of Alaska near Kodiak Island at 09:31:40.89 UTC at a depth of 14 km [25,26]. There was no significant damage reported. The earthquake woke residents in Anchorage which was located 560 km northeast from the epicenter. It was also felt in parts of British Columbia, Canada. The 2018 Mw 7.1 Anchorage earthquake happened approximately 15 km north from Anchorage, Alaska on 30 November of 2018 at 17:29:29.33 UTC at a depth of 46.7 km [9,27]. A few minutes later, a magnitude 5.8 aftershock shook the region. Significant damage has been reported to infrastructure, buildings, and airports [9]. Moreover, we investigated the characteristics of the 3 November 2002, Mw 7.9 Denali earthquake that occurred in central Alaska along a shallow strike-slip fault on the Denali-Totschunda fault system [28,29]. The details of the selected mainshocks are listed in Table 1 and the spatial distributions of the mainshocks with the corresponding aftershock sequences during the first 14 days are shown in Figure 1.

Table 1. The dates of occurrence, epicentre locations, magnitude and depth of the analyzed mainshocks.

Name	Date	Time	Latitude	Longitude	Magnitude	Depth
Denali	3 November 2002	22:12:41	63.5141	−147.4529	7.9 Mw	4.2 km
Kodiak	23 January 2018	09:31:40	56.0039	−149.1658	7.9 Mw	14 km
Anchorage	30 November 2018	17:29:29	61.3464	−149.9552	7.1 Mw	46.7 km

In this study, the left truncated exponential distribution was utilized to model the magnitude frequency statistics [30]. Moreover, the modified Omori–Utsu (OU) law [31] and Epidemic Type Aftershock Sequence (ETAS) model [32] were used to approximate the rate of the aftershocks. In addition, two statistical approaches including the extreme value distribution and Bayesian predictive distribution were utilized to compute the probabilities of having the largest expected aftershocks to be above a certain magnitude during the evolution of each sequence.

The paper is structured as follows: Section 2 begins with the specification of the earthquake catalogues and follows by defining the statistical methods to analyze the aftershock sequences. The results of the statistical analysis are provided in Section 3. In Section 4, discussion of the results and concluding remarks are given.

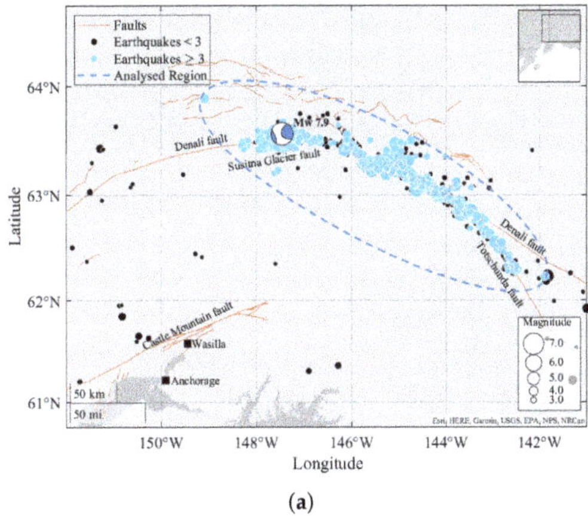

(**a**)

Figure 1. *Cont.*

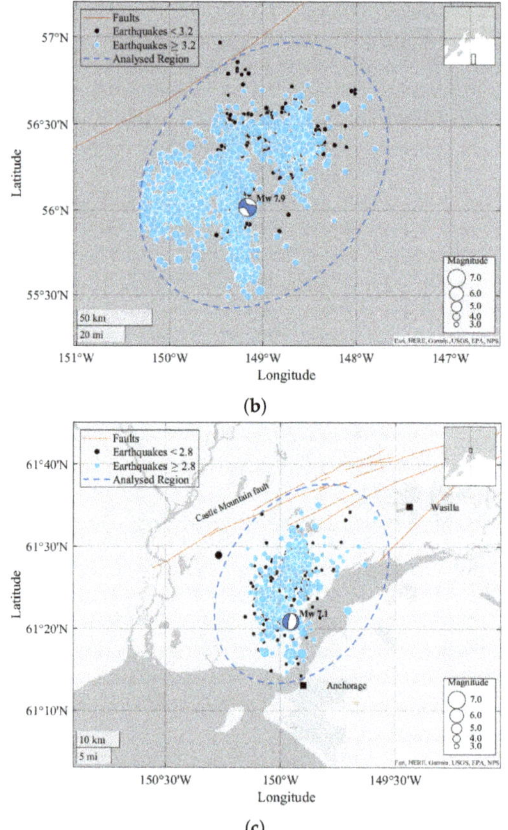

Figure 1. Maps of the occurrence of the aftershock sequences generated by the three significant Alaska mainshocks: (**a**) the 2002 Mw 7.9 Denali sequence with $m_0 = 3.0$; (**b**) the 2018, Mw 7.9 Kodiak sequence with $m_0 = 3.2$; and (**c**) the 2018 Mw 7.1 Anchorage sequence with $m_0 = 2.8$. The events during 30 days after each mainshock are plotted. The blue solid circles represent the aftershocks above m_0. Black points are all events between magnitude 2.5 and m_0. The focal mechanisms of the studied mainshocks are plotted as beach balls. Quaternary faults are plotted as light brown line segments [33].

2. Materials and Methods

2.1. Earthquake Catalogue

In order to analyze the 2002, Mw 7.9 Denali; the 2018, Mw 7.9 Kodiak, and the 2018, Mw 7.1 Anchorage earthquake sequences, the United States Geological Survey (USGS) earthquake catalogue https://earthquake.usgs.gov/earthquakes/search/ (accessed on 18 December 2021) was used. The spatial distribution of aftershocks during 30 days after each mainshock are shown in Figure 1. The focal mechanisms of the mainshocks were obtained from the USGS website [34–36].

The 2002 Denali, Alaska, earthquake sequence occurred along the Denali-Totschunda faults which is a right-lateral strike-slip fault system. The Mw 7.9 mainshock nucleated on the Susitna Glacier thrust fault and propagated further along the Denali fault and continued along the Totschunda fault [29]. The parameters of the elliptical region for this aftershock sequence, are given in Table 2 and the sequence for 30-day is depicted in Figure 1a. The plotted fault plane solution for this mainshock in Figure 1a was taken from the the USGS website [34].

The 2018 Mw 7.9 Kodiak, Alaska, earthquake took place in the Gulf of Alaska southeast of Kodiak Island. The mainshock location and the focal mechanism reflect a strike-slip faulting system within the shallow lithosphere of the Pacific plate near the subduction zone [37]. In Table 2 the details of the studied sequence for this mainshock are reported. In addition, the earthquake sequence during 30-day after the mainshock occurrence is demonstrated in Figure 1b. The focal mechanism of the 2018 Mw 7.9 Kodiak, Alaska, earthquake suggests a steeply dipping fault either as a right-lateral system that strikes the north-northwest or as a left-lateral fault that strikes west-southwest. The fault plane solution for the mainshock in Figure 1b was obtained from the USGS website [35].

The 2018 Mw 7.1 Anchorage, Alaska, earthquake happened as the result of a normal faulting rupturing north from Anchorage, Alaska. The location and mechanism of the focal mechanism reflect a moderately dipping north-south fault system fault within the subducting Pacific slab [38,39]. Details of the analyzed earthquake sequence are presented in Table 2 and the sequence of 30-day is illustrated in Figure 1c. The indicated moment tensor for this mainshock in Figure 1c was acquired from USGS [36].

Table 2. The parameters of the elliptical regions used for the identification of the aftershock sequence and the corresponding lower magnitude cutoffs m_0.

| Mainshock Name | Start Date and Time | Elliptical Aftershock Zone | | | | | | Magnitude Cut-Off, m_0 |
| | | Center | | Declination | Radii | | | |
		Latitude	Longitude		R_1	R_2		
Denali	3 November 2002 (22:12:41)	63.1	−145.40	117.5	1.85	0.55		3.0
Kodiak	23 January 2018 (09:31:40)	56.2	−149	40	0.85	0.65		3.2
Anchorage	30 November 2018 (17:29:29)	61.425	−149.91	35	0.22	0.16		2.8

For the statistical analysis of seismicity, several time intervals were utilized to estimate properly the parameters of the models describing the evolution and the statistics of the aftershock sequences. For the estimation of the model parameters, the training time interval, $[T_0, T_e]$, is considered. In order to properly account for the impact of preceding earthquakes on the earthquake rate, the training time interval is divided into an initial time interval, $[T_0, T_s]$, and a target time interval, $[T_s, T_e]$. The seismicity parameters are estimated in the target time interval. A forecasting time interval, $[T_e, T_e + \Delta T]$ is also considered to analyze the evolution and the statistics of the seismicity. The schematic illustration of the time intervals for the analysis of the aftershock sequences is shown in Figure 2.

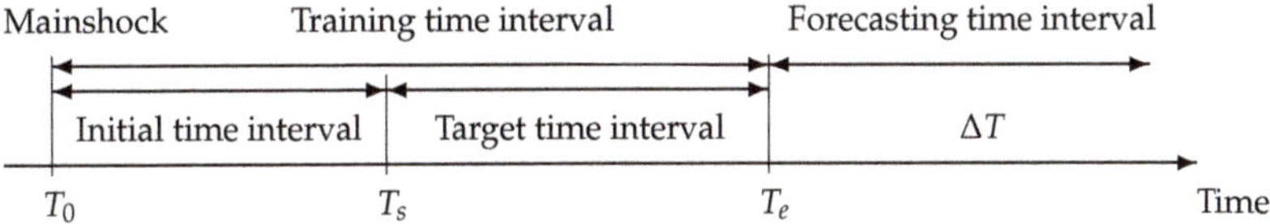

Figure 2. An illustration of the time intervals used in the analysis.

Earthquakes occur due to sudden energy release associated with the slippage of faults and are characterized by finite rupture areas. However, for statistical analysis of seismicity, the point assumption is utilized to characterize each earthquake. On time scales larger than the propagation of rupture along the fault, earthquakes can be treated as points in time and space. This idealization helps to describe the earthquake process by point process models. The point process becomes a marked point process by assigning magnitudes to each event. Therefore, each earthquake can be characterized by the magnitude, m_i, and the occurrence time, t_i, in order to generate a stochastic marked point process during a specific time interval, $S = \{(t_i, m_i)\} : i = 1, 2, \ldots, n$.

2.2. Gutenberg–Richter Scaling and the Exponential Distribution

Gutenberg and Richter [40] proposed the relationship that describes the frequency-magnitude statistics of earthquakes. This relationship between the frequency of event occurrences and the event magnitudes is one of the most commonly used empirical laws in statistical seismology. They suggested the following equation:

$$N(m \geq) = 10^{(a-bm)}, \tag{1}$$

where $N(m \geq)$ is the total number of earthquakes above magnitude m and $N(0 \geq) = 10^a$ and b is the value of the slope of the fitted line to the N on a logarithmic scale. Vere-Jones [30] emphasized that the distribution of earthquake magnitudes is described by the exponential distribution for $m \geq m_0$ with the probability density, $f_\theta(m)$, and cumulative distribution function, $F_\theta(m)$:

$$f_\theta(m) = \beta e^{-\beta(m-m_0)}, \tag{2}$$

$$F_\theta(m) = 1 - e^{-\beta(m-m_0)}, \tag{3}$$

where m_0, is the lower magnitude cut-off that is above the catalogue completeness magnitude m_c and $\theta = \{\beta\}$ is the model parameter which can be obtained from all earthquakes above m_0 in target time interval $[T_s, T_e]$. The parameter β is related to the b-value of the Gutenberg–Richter (GR) scaling:

$$\beta = b \ln(10). \tag{4}$$

The Maximum Likelihood Estimation (MLE) is the most common approach to estimate the b-value or parameter β. Bender [41] suggested an estimator for β by taking into account the binning of the magnitude. Tinti and Mulargia [42] proposed an approach to calculate the uncertainties of the parameter β at a given confidence level.

2.3. Omori–Utsu Law

For the first time, Omori [43] introduced a formula for the aftershock sequence decay rate, $\lambda(t)$, that is inversely related to the elapsed time after the mainshock. Utsu [31] proposed a modification of the Omori law which is known as the Omori–Utsu (OU) law. Utsu [31] modified the original intensity to the following form:

$$\lambda_\omega(t) = \frac{K}{(t+c)^p}, \tag{5}$$

where $\lambda_\omega(t)$ is the earthquake rate at a given time t with magnitudes above m_0, and set of parameters $\omega = \{K, c, p\}$, and t is the time elapsed since the occurrence of the mainshock at $T_0 = 0$. The parameter K is the productivity of the sequence, c is the characteristic time, and p is the rate of the decay in time. By considering the non-homogeneous Poisson process for the occurrence of earthquakes, the parameters $\omega = \{K, c, p\}$ can be determined by using the MLE approach [44,45]. In addition, in this model, the parameter uncertainties can be estimated from the inverse of the Fisher information matrix that is computed from the likelihood function.

2.4. The Epidemic Type Aftershock Sequence (ETAS) Model

A more realistic approximation of the earthquake rate was proposed by Ogata [46], where he suggested that each earthquake could be considered as a trigger for the next events in the sequence. The conditional intensity of the temporal ETAS model, $\lambda_\omega(t|\mathcal{H}_t)$, at time t is defined as [46]:

$$\lambda_\omega(t|\mathcal{H}_t) = \mu + A \sum_{i:t_i<t} \frac{e^{\alpha(m_i-m_0)}}{(\frac{t-t_i}{c}+1)^p}, \tag{6}$$

where $\omega = \{\mu, A, c, p, \alpha\}$ is the set of parameters of temporal conditional intensity with a reference magnitude m_0 and the occurrence history of earthquakes, $\mathcal{H}_t$, during the time interval $[T_0, t]$. N_t is the number of the earthquakes with magnitudes above m_0 in the time interval $[T_0, t]$. In the ETAS model, μ specifies the average rate of background events that transpire independently of any other events. c is the temporal characteristic time, p governs the rate of decay of triggered events as a power law, and A controls the event productivity. The parameter α determines the degree of aftershock clustering. Larger values of α correspond to more pronounced aftershock sequences with stronger variability in earthquake magnitudes. In contrast, the impact of event's magnitude on aftershock generation is reduced by smaller α values. The estimation of the ETAS model parameters is achieved by maximizing the log-likelihood function:

$$\log L = \sum_{i:t_i \leq T_e} \lambda_\omega(t_i | \mathcal{H}_{t_i}) - \int_{T_s}^{T_e} \lambda_\omega(t | \mathcal{H}_t)\, dt. \tag{7}$$

In general, the consistency of the ETAS model is measured on the basis of a transformed time. The transformed time τ_i for a given event is computed by using the cumulative conditional intensity at time t_i as

$$\tau_i = \int_0^{t_i} \lambda_\omega(t)\, dt. \tag{8}$$

If the fit of the model is accurate, the sequence of earthquakes should obey a stationary Poisson process in the transformed time. Furthermore, the cumulative number of observed earthquakes in transformed time can be close to a straight line [13]. The deviation of the cumulative number of observed events from the straight line indicates that the model does not fit well the earthquake sequence.

2.5. Extreme Value Distribution

By considering a non-homogeneous Poisson sequence of earthquakes, the probability of having an extreme earthquake with a magnitude above m in the forecasting time interval, $[T_e, T_e + \Delta T]$ can be obtained from the Extreme Value Distribution (EVD) [47]:

$$\Pr_{EV}\{m_{ex} \geq m | \theta, \omega, \Delta T\} = 1 - e^{-\{\Lambda_\omega(\Delta T)[1 - F_\theta(m)]\}}, \tag{9}$$

where m_{ex} is the magnitude of the largest expected event, ω is the set of parameters of seismicity rate $\lambda_\omega(t)$, $F_\theta(m)$ is the cumulative distribution function of the events' magnitude with the set of parameters θ, and $\Lambda_\omega(\Delta t)$ is a productivity function that is given as:

$$\Lambda_\omega(\Delta T) = \int_{T_e}^{T_e + \Delta T} \lambda_\omega(t)\, dt. \tag{10}$$

By considering the exponential model, Equation (3), for describing the magnitude distribution and the UO model, Equation (5), for the intensity of the productivity function, Equation (9) can be rewritten as:

$$\Pr_{EV}\{m_{ex} \geq m \mid \theta, \omega, \Delta T\} =$$
$$1 - \exp\left\{-\left[K\frac{(T_e + c)^{1-p} - (T_e + \Delta T + c)^{1-p}}{p - 1}\right](e^{-\beta(m - m_0)})\right\}, \tag{11}$$

for $p \neq 1$, and the set of parameters $\{\theta, \omega\}$ can be obtained during the target time interval $[T_s, T_e]$. Therefore from Equation (11), the probability of having an earthquake with a magnitude above m in a forecast time interval $[T_e, T_e + \Delta T]$ can be obtained, which is the same approach as in Reasenberg and Jones [8].

2.6. Bayesian Predictive Distribution

The obtained parameters of the aftershock sequence model during the training time interval play a crucial role in calculating the EVD. The uncertainty of the parameters have a significant impact on the calculation of the corresponding probabilities. Shcherbakov et al. [3,48] incorporated the model uncertainties into the computation of the probabilities for the occurrence of the largest expected earthquakes by applying the Bayesian predictive distribution (BPD) approach, in which the BPD can be defined as:

$$\mathrm{Pr}_B\{m_{ex} \geq m \mid S, \Delta t\} = \int_\Omega \int_\Theta \mathrm{Pr}_{EV}(m_{ex} \geq m \mid \theta, \omega, \Delta T)p(\theta, \omega \mid S)\, d\theta d\omega, \quad (12)$$

where Θ and Ω are the frequency-magnitude distribution and seismicity rate parameter domains, respectively. $\mathrm{Pr}_{EV}(m_{ex} \geq m \mid \theta, \omega, \Delta T)$ is the EVD and $p(\theta, \omega \mid S)$ is the posterior distribution function, which quantifies the uncertainties of the model parameters.

Since the ETAS model deviates from a non-homogeneous Poisson process the EVD for the largest magnitudes is not given by Equation (9). Shcherbakov et al. [3] suggested to use the stochastic simulations to approximate the extreme value distribution and ultimately the BPD. In this approach, the Metropolis-within-Gibbs algorithm is used to sample from the conditional posterior distribution to generate the chain of the model parameters using the Markov Chain Monte Carlo (MCMC), then the model parameter chain is used to simulate the ETAS model during the forecasting time interval $[T_e, T_e + \Delta T]$. At the end, the maximum magnitude is taken from each sequence of events to construct a distribution that approximates the BPD.

When performing MCMC sampling a certain initial part of the parameter chain is discarded as "burn-in". The Gamma distribution was considered for the prior distribution of the model parameters. As burn-in the first 50% of Markov chains were discarded and the second half was utilized for calculation of the BPD.

2.7. Forecast Validation

To evaluate the number of forecasted earthquakes by a specific model in the forecasting time interval, the N-test can be used [4,17,19,49]. It tests the distribution range of the number of the forecasted events versus the number of observed earthquakes. In addition, in order to test the magnitude distribution of the forecasted earthquakes the M-test can be applied [4,17,19,49]. The N and M-tests examine the consistency of the forecasts with respect to observations, and the R-test can be used to compare the performance of different forecasting models [17]. In addition, to evaluate statistical forecast the T-test can be applied [49]. In formulating the T-test, the sample information gain per earthquake of the model Λ^2 over the model Λ^1 is defined as $I_N(\Lambda^2, \Lambda^1) = R^{21}/N_{\text{obs}}$, where N_{obs} is the number of observed earthquakes during the forecasting time interval ΔT and $R^{21} = L(\mathrm{M}|\Lambda^2) - L(\mathrm{M}|\Lambda^1)$ is the log-likelihood ratio of the two models. The detailed explanation and implementation of these tests applied to the time-dependent models such as the ETAS and OU rates can be found in Shcherbakov [4].

3. Results

In this section the obtained results of the analysis of the recent three significant mainshocks that occurred in Alaska (the 2002, Mw 7.9 Denali; the 2018, Mw 7.9 Kodiak, and the 2018; Mw 7.1 Anchorage earthquakes) are summarized.

3.1. Frequency-Magnitude Statistics Analysis

The aftershocks of the three mainshocks within elliptical regions as shown in Figure 1 were used to obtain the frequency-magnitude statistics. The fitting of Equation (2) was done to all three sequences and during specific target time intervals. To estimate the parameter β from Equation (2) the MLE approach was used [41]. The model parameter uncertainties were estimated using the method of Tinti and Mulargia [42]. In addition, the method of goodness of fit test [50] was utilized to estimate the magnitude of completeness m_c for the three sequences. Specifically, m_0 was selected as the magnitude above which at least 95% of the observed data are modeled by Equation (2). The results are presented in Figure 3. Moreover, to investigate how the earthquake magnitudes evolve over time, they are plotted versus the sequential number in Figure A1. This can be used to inspect the early magnitude incompleteness of aftershock sequences and can be used to justify the use of a chosen magnitude threshold [51].

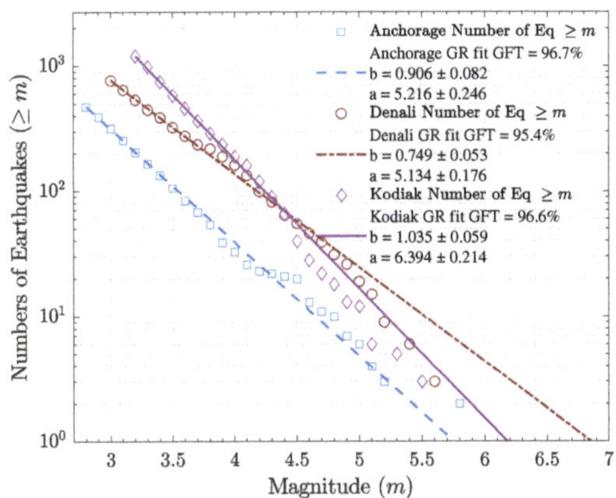

Figure 3. The frequency-magnitude statistics of aftershock sequences for 30 days from the mainshock occurrence. The lines are the Gutenberg–Richter scaling fit, Equation (1). The open symbols represent the cumulative numbers corresponding to each aftershock sequence. The estimated a and b-value with 95% confidence intervals are given in the legend. The cumulative numbers of aftershocks for the 2002, Denali aftershocks for $m \geq 3.0$ are plotted as red circles, the 2018, Kodiak aftershocks for $m \geq 3.2$ cumulative numbers are plotted as purple diamonds, and blue squares are used to depict the 2018, Anchorage aftershocks for $m \geq 2.8$.

In order to analyze the frequency-magnitude statistics of the 2002, Denali earthquake sequence, $m_0 = 3.0$ was considered as a cut-off magnitude, and the analysis was performed during $[T_s, T_e] = [0, 30]$ days after the mainshock occurrence on 3 November 2002 (22:12:41 UTC) for the earthquakes within the elliptical region given in Figure 1a. The total number of aftershocks during the selected time interval was 771 with the maximum magnitude 5.6, respectively. The fit of the GR relation is demonstrated in Figure 3. The estimated b-value and a-value for the analyzed earthquake sequence are 0.749 ± 0.053 and 5.134 ± 0.176, respectively. The magnitude-frequency statistics analysis of the 2018, Kodiak earthquake sequence was performed during $[T_s, T_e] = [0, 30]$ days after the mainshock that occurred on 23 January 2018 (09:31:4 UTC) for earthquakes with the cut-off magnitude $m_0 = 3.2$ within an elliptical region shown in Figure 1b. In total 1207 earthquakes with the magnitude ranging from 3.2 to 5.5 occurred in the analyzed elliptical region during the specified time interval. The estimated b-value and a-value for selected earthquake sequence are 1.035 ± 0.059 and 6.394 ± 0.214, respectively (Figure 3). For analyzing the frequency-magnitude statistics of the 2018, Anchorage earthquake sequence, $m_0 = 2.8$ was considered as a cut-off magnitude, and the analysis was carried out during $[T_s, T_e] = [0, 30]$ days after the mainshock occurrence on 30 November 2018 (17:29:29 UTC) for the earthquakes within an elliptical region of Figure 1c. In total 476 earthquakes within the magnitude range of 2.8–5.2 occurred in the analyzed spatiotemporal window. The obtained b-value and a-value for Anchorage earthquake sequence are 0.906 ± 0.082 and 5.216 ± 0.246 respectively. The fit of the GR relation is plotted in Figure 3.

3.2. Aftershock Decay Rate Modelling

To begin with, we obtained the first-order approximation to the background seismicity rate within the presented elliptical regions in Figure 1 for each analyzed aftershock sequence. In this evaluation the background rate was estimated as the ratio of the number of earthquakes to the number of days, during the time interval that started on 1 January 2000, and ended 30 days before each mainshock. The estimated background rates, the corresponding time intervals, and cut-off magnitude for each analyzed mainshock are reported in Table 3.

Table 3. Background Seismicity Rate.

Name	Start Time	End Time	m_0	μ (Events per Day)
Denali	1 January 2000	3 October 2002	3.0	0.1
Kodiak	1 January 2000	23 December 2017	3.2	0.005
Anchorage	1 January 2000	30 October 2018	2.8	0.02

Subsequently, the aftershock decay rate was modeled by using the OU law, Equation (5), for the three sequences during the target time interval of $[T_s, T_e] = [0.001, 30]$ days. The obtained parameters of the OU model with 95% confidence interval and the model fits are shown for the Anchorage sequence in Figure 4 and for the Denali and Kodiak sequences in Figure A2.

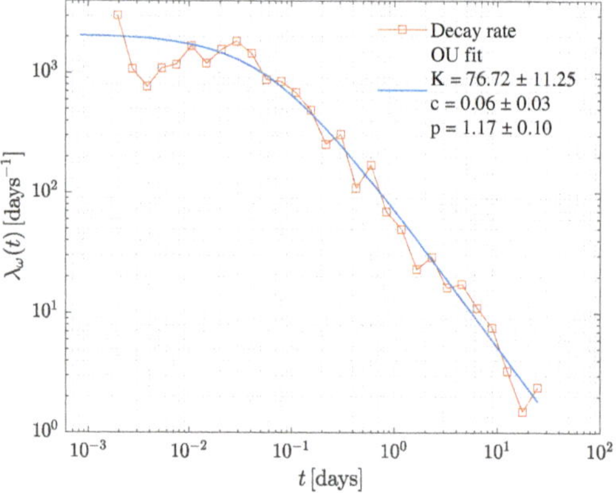

Figure 4. The log-log plot of the aftershock decay rate for the 2018, Mw 7.1 Anchorage aftershock sequence with magnitudes $m \geq 2.8$ are presented as open squares. The blue solid line is the corresponding fit of the OU law, Equation (5), to the aftershock sequence. The obtained parameters from the OU law, Equation (5), with the 95% confidence intervals are reported in the legend.

The estimated set of parameters of the OU law, Equation (5), with 95% confidence intervals for the 2002, Denali earthquake sequence are $\omega = \{K = 202.72 \pm 48.35, c = 0.29 \pm 0.13, p = 1.22 \pm 0.11\}$ (Figure A2a). Similarly for the 2018, Kodiak earthquake sequence the obtained parameters of the OU law, Equation (5) with 95% confidence intervals are $\omega = \{K = 225.91 \pm 46.84, c = 0.31 \pm 0.18, p = 0.88 \pm 0.09\}$ (Figure A2b). Finally, for the 2018, Anchorage earthquake sequence the estimated parameters of the OU law are $\omega = \{K, c, p\}, \{76.71 \pm 11.25, 0.06 \pm 0.03, 1.17 \pm 0.1\}$, respectively. The fit of the OU law and the estimated parameters from Equation (5) are demonstrated in Figure 4.

Furthermore, we used the ETAS model, Equation (6), to estimate the aftershock decay rate during the target time interval $[T_s, T_e] = [0.06, 30]$ days for the three aftershock sequences. For comparison we present the OU fit and the ETAS model fit for the 2018, Anchorage earthquake sequence in Figure 5. A similar plot for the other two sequences is given in Figure A3. Figure 6 illustrates the fit of the ETAS model in transformed time for the 2018, Anchorage earthquake sequence.

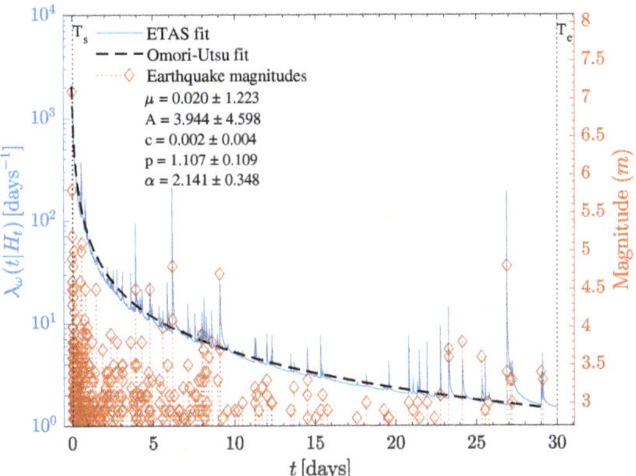

Figure 5. The aftershock sequence and corresponding earthquake magnitudes during 2018, Mw 7.1 Anchorage sequence with $m \geq 2.8$. The ETAS model fit, Equation (6), for the target time interval of $[T_s, T_e] = [0.06, 30]$ is plotted as a solid blue line, and the obtained set of parameters are reported with 95% confidence intervals. The OU law fit, Equation (5), is plotted as a black dashed line for comparison.

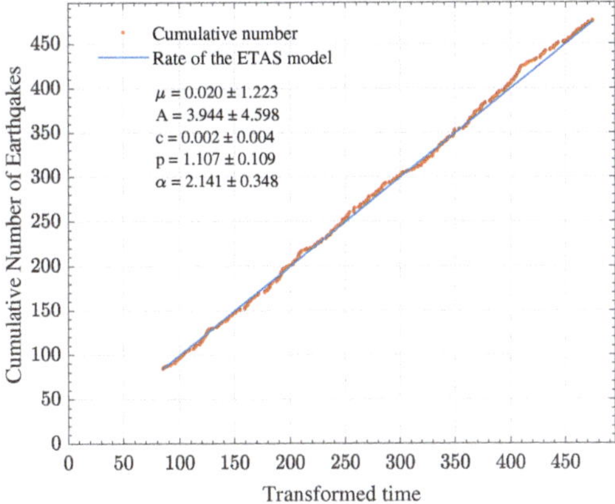

Figure 6. The cumulative number of observed aftershocks in transformed time and corresponding rate of the ETAS model for the 2018, Mw 7.1 Anchorage aftershock sequence with $m \geq 2.8$. The ETAS model fit, Equation (6), for the target time interval of $[T_s, T_e] = [0.06, 30]$ days is plotted as a solid blue line, and the obtained set of parameters are reported with 95% confidence intervals.

Finally, the statistical properties of the aftershock sequence initiated by the M7.1 Anchorage, Alaska, earthquake are investigated in detail during several additional target time intervals. Specifically, the sequence was analyzed during several target time intervals starting from the occurrence of the mainshock and ending at $T_e = [1, 2, 3, 4, 5, 6, 7, 10, 14, 21, 30]$ days. The evolution of the estimated parameters with 95% confidence intervals for both models OU and ETAS during 2018, Mw 7.1 Anchorage earthquake sequences are shown in Figure A4. Obtained estimations for the b-value of the GR relation, Equation (1) with 95% confidence intervals are demonstrated in Figure A4a. The evolution of the OU model parameters are shown in Figure A4b. We presented the evolution of the estimation of the ETAS model parameters in Figure A4c.

3.3. Retrospective Forecasting of the Largest Expected Aftershocks

In order to calculate the probability of having the magnitude of the largest expected aftershock to be above a certain magnitude and during a predefined forecasting time interval the EVD, Equation (9), and BPD, Equation (12), are used. In this analysis, the OU law, Equation (5), and the ETAS model, Equation (6), are utilized to calculate the aftershock decay rate, and the frequency-magnitude distribution is estimated from the exponential distribution, Equation (3).

To illustrate the applicability of the methods, one particular example is illustrated in case of the 2018 Anchorage sequence. The training time interval was set to $[T_s, T_e] = [0.06, 14]$ days and the forecasting time interval of $\Delta T = 7$ days was considered. The lower magnitude cut-off $m_0 = 2.8$ was used. The computed distributions using the EVD, Equation (9), and BPD, Equation (12) are plotted in Figure 7. For the BPD analysis, total of 20,000 MCMC sampling of the posterior distribution was performed. The first 10,000 iterations were discarded as "burn-in" and the remaining 10,000 samples were utilized to perform stochastic simulations of the ETAS or OU processes. The resulting distributions of the OU and ETAS model parameters estimated from the MCMC chains are reported in Table A1 and plotted in Figures A5 and A6, respectively.

Moreover, two cases are considered for computing the probabilities for the occurrence of the largest expected aftershocks above a certain magnitude during the evolution of the three sequences. For the first case, we considered a constant forecasting time interval $\Delta T = 7$ days. As for the target time intervals, we considered the following ending times $T_e = [1, 2, 3, 4, 5, 6, 7, 10, 14, 21, 30]$ days with the lower magnitude thresholds of 3.0, 3.2, and 2.8 for the 2002, Denali, 2018, Kodiak, and 2018, Anchorage sequences, respectively. In this analysis, the BPD, Equation (12), with the exponential distribution, Equation (3), for the frequency magnitude statistics, and the ETAS model, Equation (6), for the occurrence rate of the earthquakes were utilized. The obtained results for the probabilities of the largest expected earthquakes greater than $m_{\mathrm{ex}} \geq 5.0, 6.0, 7.0$ are shown in Figure 8. Furthermore, the probabilities of having the largest expected aftershock to be above magnitude 6.0 were computed utilizing the EVD, Equation (8), combined with the OU law, Equation (5), and using the BPD, Equation (12), combined with the OU law, Equation (5), or the ETAS model, Equation (6). The obtained results for analyzed mainshocks are presented in Figure 9.

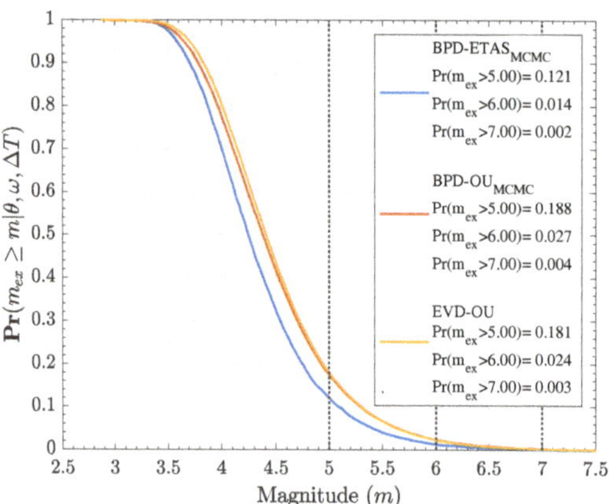

Figure 7. The EVD and BPD for the 2018, Mw 7.1 Anchorage aftershock sequence during the 7-day forecast time interval after the training time interval of 14 days. The blue solid line represents the BPD using the ETAS model with 10000 MCMC sampling steps using the Gamma prior. The orange line represents the obtained BPD using the OU model and the yellow line is the plot of the EVD with the OU law, Equation (11).

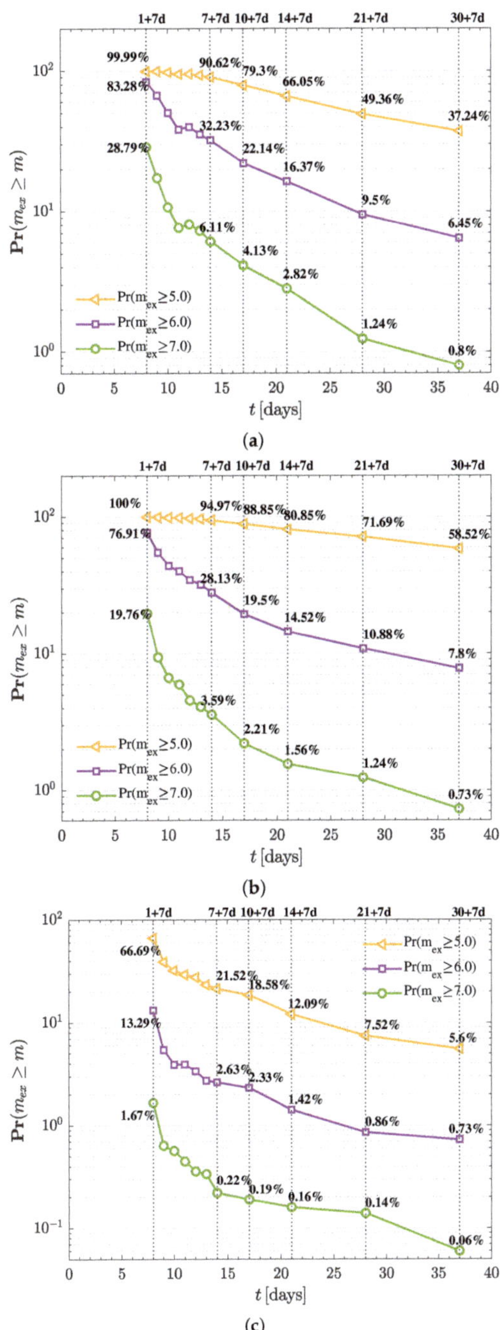

Figure 8. The probabilities to have the largest expected aftershocks to be larger than $m_{ex} \geq 5.0,\ 6.0,\ 7.0$ using the BPD, Equation (12), during a constant forecasting time interval $\Delta T = 7$ days and for the varying target time intervals. (**a**) The 3 November 2002, Mw 7.9 Denali sequence with $m \geq 3.0$. (**b**) The 23 January 2018, Mw 7.9 Kodiak sequence with $m \geq 3.2$. (**c**) The 30 November 2018, Mw 7.1 Anchorage sequence with $m \geq 2.8$.

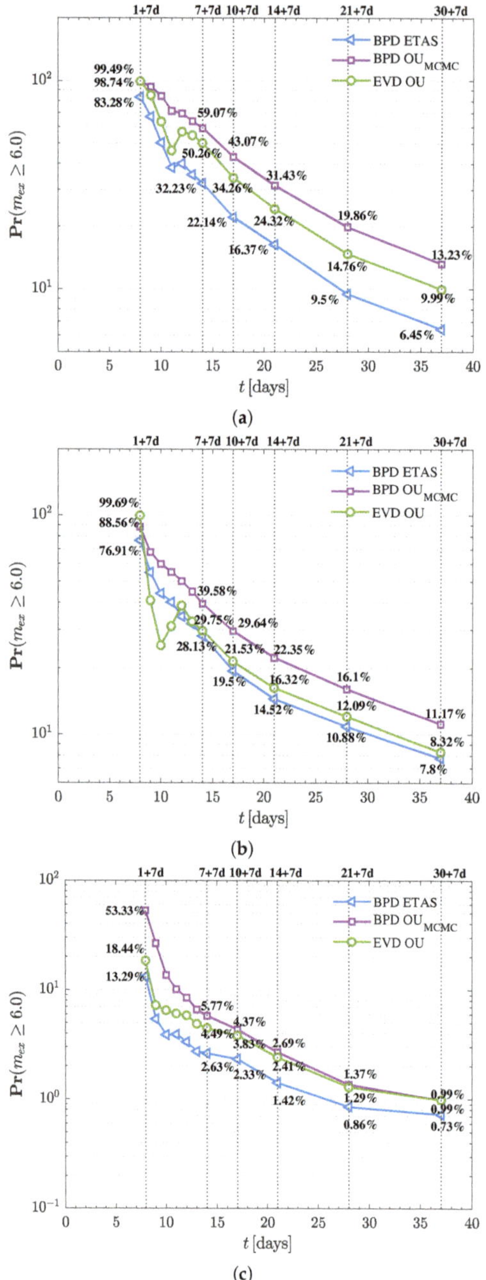

Figure 9. The comparison of the probabilities to have the largest expected aftershock during the forecasting time interval $\Delta T = 7$ days for the three sequences: (**a**) the 3 November 2002, Mw 7.9 Denali for $m \geq 3.0$; (**b**) the 23 January 2018, Mw 7.9 Kodiak for $m \geq 3.2$; (**c**) the 30 November 2018, Mw 7.1 Anchorage for $m \geq 2.8$. The blue triangles are computed using the BPD, Equation (12), with an earthquake decay rate given by the ETAS model, Equation (6). The purple squares are computed using BPD, Equation (12), with an earthquake decay rate given by OU law, Equation (5). The green circles give probabilities computed using the EVD, Equation (11).

In the second case, a constant target time interval $[T_s, T_e] = [0.06, 2]$ days was considered. However, the forecasting time interval was varied as $\Delta T = [1, 2, 5, 7, 10, 14]$ days to compute the probabilities of the occurrence of the largest expected aftershocks. The computed probabilities for the largest anticipated earthquakes $m_{ex} \geq 5.0, 6.0, 7.0$ are illustrated in Figure A7. In addition, the comparison of the two approaches to compute the probabilities (EVD versus BPD) combined with either the OU law or the ETAS model are shown in Figure A8.

3.4. Testing the Model Forecasts

Several tests were conducted to evaluate the forecast during the time interval $[T_e, T_e + \Delta T]$ by comparing the simulated results with the observed seismicity. To check the performance of forecasts for the number of aftershocks and magnitude distribution, the N and M-tests were performed, respectively. The details of the implementation of the tests can be found in Shcherbakov [4].

For the three aftershock sequences the number of forecasted aftershocks in the forecasting time interval $\Delta T = 7$ days and using the following target time intervals $T_e = [1, 2, 3, 4, 5, 6, 7, 10, 14, 21, 30]$ days are given in Figure 10. For comparison, in the same figure the observed number of earthquakes are shown as blue circles for each prediction time interval ΔT. In addition, for the constant target time interval $[T_s, T_e] = [0.06, 2]$ days and varying forecasting time intervals $\Delta T = [1, 2, 5, 7, 10, 14]$ days the number of forecasted and observed earthquakes are shown in Figure A9. To investigate the effect of the magnitude cutoff, we also performed the same analysis for earthquakes above magnitude 3.5. This is reported in Figure A10.

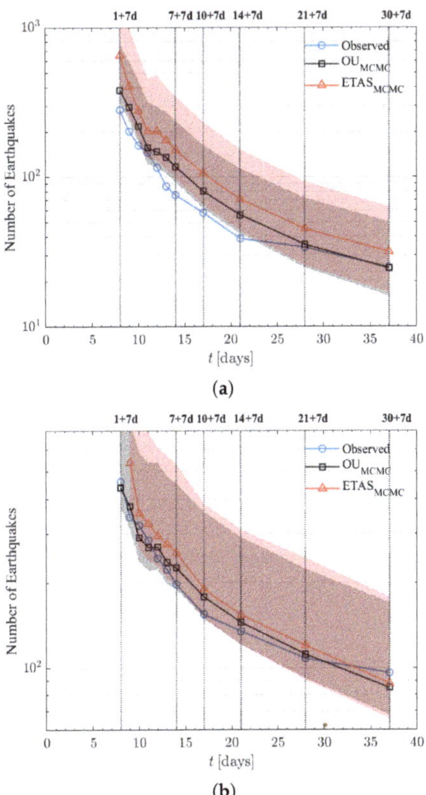

Figure 10. *Cont.*

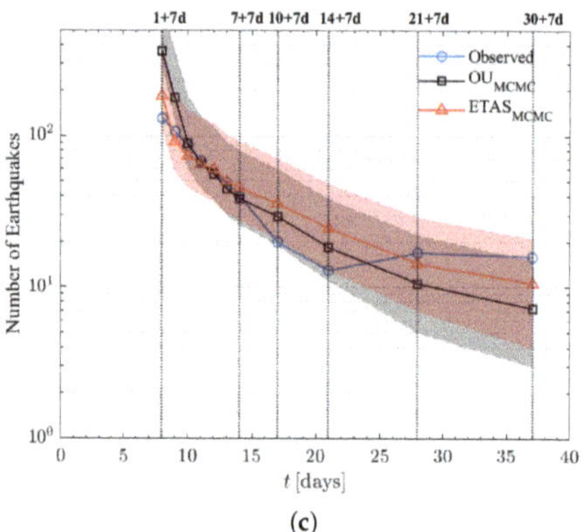

(c)

Figure 10. The number of forecasted and observed aftershocks in the forecasting time interval $\Delta T = 7$ days for the three sequences: (**a**) the 3 November 2002, Mw 7.9 Denali sequence for $m \geq 3.0$; (**b**) the 23 January 2018, Mw 7.9 Kodiak sequence for $m \geq 3.2$; (**c**) the 30 November 2018, Mw 7.1 Anchorage sequence for $m \geq 2.8$. The red triangles show the average number of forecasted earthquakes using the ETAS model and the black squares illustrate the average number of forecasted earthquakes using OU law. The shading bands represent 95% confidence intervals. The blue circles represent the observed number of earthquakes in each forecasting time interval.

In addition, M-test was performed to assess the consistency of the distribution of the magnitudes of the forecasted events. The results of the performance of the OU law and ETAS model are reported by computing the quantile score, κ [4,19]. κ is defined as the proportion of the forecasted magnitudes compared to the observed magnitudes in each magnitude bin. The obtained quantile scores for the constant forecasting time interval $\Delta T = 7$ days are given in Figure 11. In addition, in Figure A11 the outcomes for the M-test for the constant target time interval $[T_s, T_e] = [0.06, 2]$ days are plotted for the three aftershock sequences.

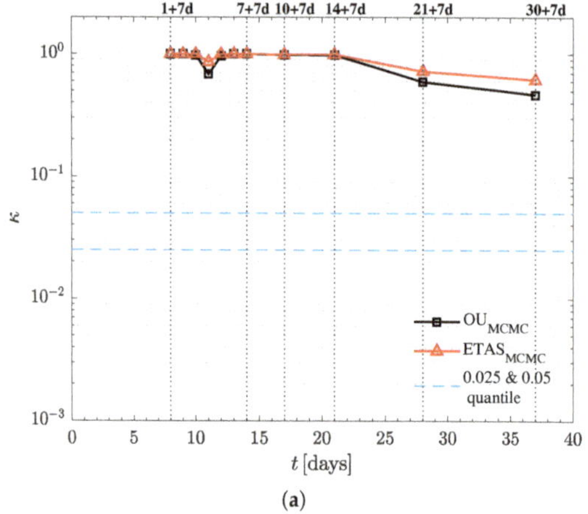

(a)

Figure 11. *Cont.*

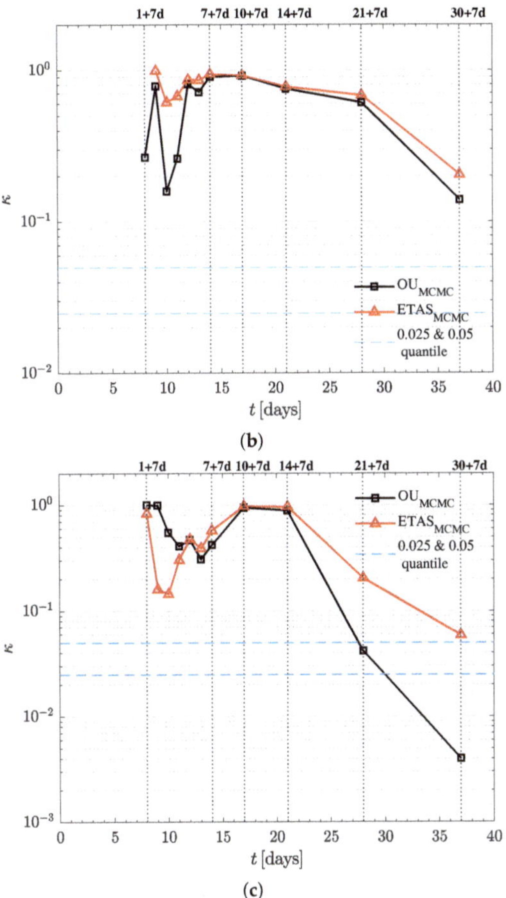

(b)

(c)

Figure 11. The obtained quantile scores from the M-test for the constant forecasting time interval $\Delta T = 7$ days for the three sequences: (**a**) the 3 November 2002, Mw 7.9 Denali for $m \geq 3.0$; (**b**) the 23 January 2018, Mw 7.9 Kodiak for $m \geq 3.2$; (**c**) the 30 November 2018, Mw 7.1 Anchorage for $m \geq 2.8$. The red triangles demonstrate the obtained quantile scores from the ETAS model and the black squares illustrate the quantile scores of OU law. The blue dashed lines represent the 0.025th and 0.05th quantiles.

In order to evaluate and compare the models, the R-test and T-test were applied for both cases by considering the ETAS model versus the OU law. In the R-test the quantile score, α, was calculated. α is the proportion of the simulated likelihood ratios, over the observed likelihood ratios [17]. The values of α that are greater than a specific level of significance support the model that was chosen as a base model, in this case it is the ETAS model. The obtained result for the α from the OU law versus the ETAS model is shown in Figure 12. Furthermore, in Figure 13, the ratio of the likelihood score of the ETAS model and the OU law over the number of the observed events in the forecasting time interval is given. The T-test is used to assess whether the sample information gain is statistically different from zero. This is used to select the preferred model [49]. Lastly, the Bayesian p-value of the BPD analysis using either the ETAS model or the OU law are illustrated for both cases in Figure 14.

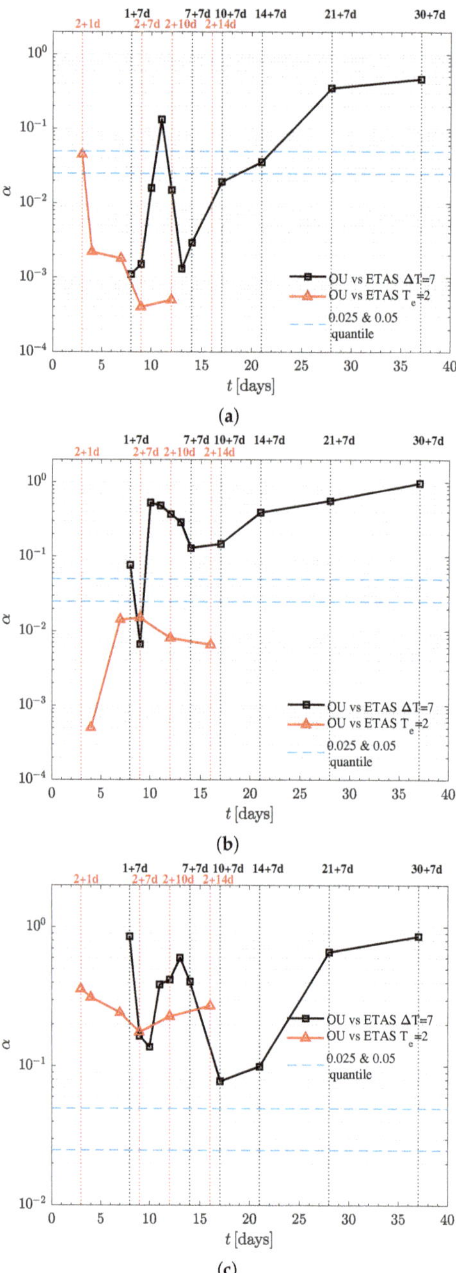

Figure 12. The obtained quantile scores from the R-test to compare the forecast based on the OU law versus the ETAS model for the three sequences: (**a**) the 3 November 2002, Mw 7.9 Denali for $m \geq 3.0$; (**b**) the 23 January 2018, Mw 7.9 Kodiak for $m \geq 3.2$; (**c**) the 30 November 2018, Mw 7.1 Anchorage for $m \geq 2.8$. The black squares illustrate the quantile scores in the case with the constant forecasting time interval of $\Delta T = 7$ days. The red triangles show the obtained quantile scores from the second case with the constant target time interval of $[T_s, T_e] = [0.06, 2]$ days. The blue dashed lines represent the 0.025th and 0.05th quantiles.

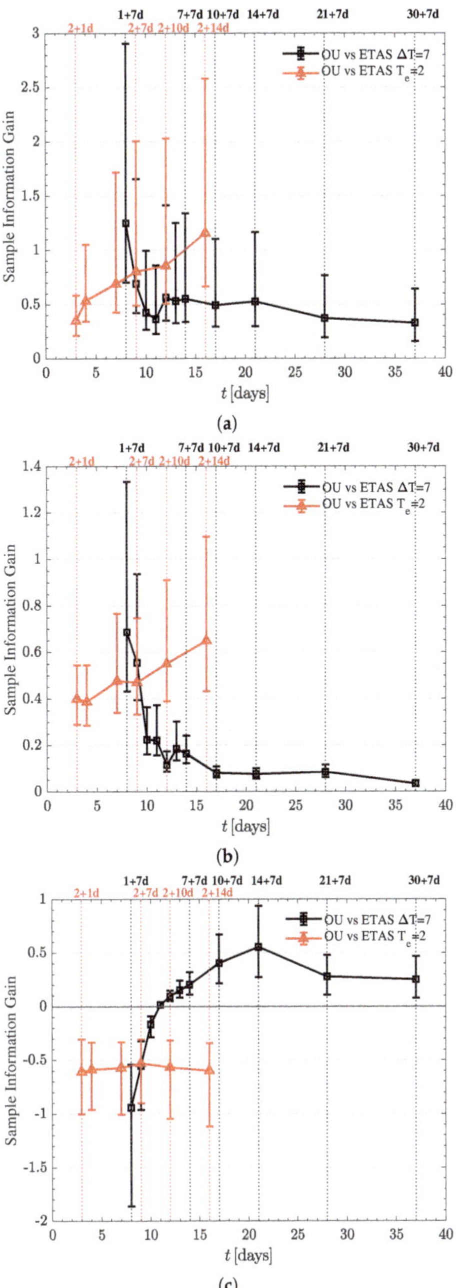

Figure 13. The sample information gain of the ETAS model versus the OU law over the number of observed data for the three sequences: (**a**) the 3 November 2002, Mw 7.9 Denali for $m \geq 3.0$; (**b**) the 23 January 2018, Mw 7.9 Kodiak for $m \geq 3.2$; (**c**) the 30 November 2018, Mw 7.1 Anchorage for $m \geq 2.8$. The black solid squares illustrate the sample information gain for the first case with the fixed the forecasting time interval of $\Delta T = 7$ days. The red triangles demonstrate the sample information gain for the second case for the constant target time interval $[T_s, T_e] = [0.06, 2]$ days.

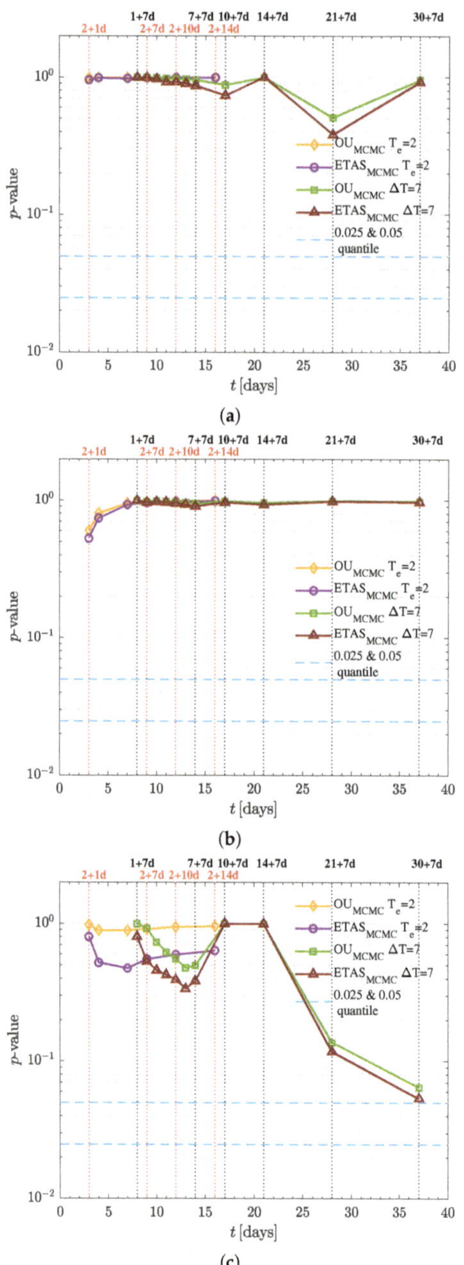

Figure 14. For both cases the Bayesian p-value of the BPD from the ETAS model and the OU law are illustrated for the three sequences: (**a**) the 3 November 2002, Mw 7.9 Denali for $m \geq 3.0$; (**b**) the 23 January 2018, Mw 7.9 Kodiak for $m \geq 3.2$; (**c**) the 30 November 2018, Mw 7.1 Anchorage for $m \geq 2.8$. The green squares and the red triangles illustrate the obtained p-value from the first case, with the fixed forecasting time interval $\Delta T = 7$ days. The yellow diamonds and purple circles demonstrate the p-value for the second case with the constant target time interval of $[T_s, T_e] = [0.06, 2]$ days. The blue dashed lines represent the 0.025th and 0.05th quantiles.

4. Discussion and Conclusions

To describe the three aftershock sequences which occurred in the Alaska region, statistical models were used in this study. To be more precise, the frequency-magnitude statistical analysis was performed using the GR relation, and the occurrence rates of the aftershock sequence were estimated by the OU law and the ETAS model. The EVD and BPD approaches were used to calculate the probability of having the largest expected aftershock above a certain magnitude during evolution of each sequence for various training and forecasting time intervals.

The frequency-magnitude distributions and estimated GR parameters for analyzed sequences are shown in Figure 3. The frequency-magnitude distribution of the Denali aftershock sequence, was characterized by a broad distribution of event magnitudes which led to a relatively low b-value (0.749 ± 0.053) (Figure 3). The 2002, Denali mainshock was followed by several large aftershocks with the largest being 5.6 magnitude which occurred in the first 24 h after the mainshock. The frequency-magnitude statistical analysis of the aftershock sequence of the Kodiak 7.9 magnitude earthquake with the cut-off magnitude 3.2 for the target time interval $[T_s, T_e] = [0, 30]$ days was performed (Figure 3). The frequency-magnitude distribution of the 2018, Kodiak sequence indicates a typical GR fit with b-value (1.035 ± 0.059). The mainshock was followed by several large aftershocks with the largest being 5.5 magnitude event. It should be noted the epicenter of the 2018, Kodiak earthquake was located in a remote area in the North Pacific. The analysis of the 2018 Anchorage sequence produced the b-value of 0.906 ± 0.082 (Figure 3). The largest aftershock with a magnitude 5.8, was close to the expected magnitude from Båth's law [5] which states that the largest aftershock is on average 1.2 magnitudes lower than the mainshock.

In order to analyze the occurrence rate of the aftershock sequence of the selected mainshocks, the Omori–Utsu law, Equation (5), and the ETAS model, Equation (6), were utilized. The obtained results from the analysis of the decay rate of the aftershock sequences show that the parameter p is comparable for both models (the OU law and the ETAS model) except for the 2018 Kodiak sequence.

Computing the probability of having a largest expected aftershock with a magnitude above a given value during different forecasting time intervals after the mainshock was one of the main objectives of this work. The EVD and BPD approaches were used to accomplish this objective.

The obtained result of this analysis indicates that the BPD method using the ETAS model is more conservative than BPD using the OU law and the EVD approach. In addition, for the aforementioned approaches, the probabilities of having an earthquake with magnitude 6 and above were calculated for both cases (Figures 9 and A8).

Moreover, in order to compare characteristics of analyzed sequences the probabilities to have the largest expected aftershocks to be larger than $m_{ex} \geq 5.0, 6.0, 7.0$ were estimated by using the BPD, Equation (12), for both cases (Figures 8 and A7). The results of this analysis show that the Anchorage sequence had a lower potential to generate aftershocks with $m_{ex} \geq 5.0, 6.0, 7.0$ compared to other analyzed sequences. These statistical results can be explained directly by the number of events in the aftershock sequence and the magnitude of the mainshock. This increases the probability of occurrence of an aftershock with a certain magnitude in a predefined time interval after the mainshock. In the present implementation of the EVD and BPD analysis we used the unbounded GR distribution. It was suggested that more realistic truncated magnitude distribution can be more appropriate for forecasting [52,53]. This can be easily incorporated in the analysis as well.

The N-tests, M-test, R-test, and T-test were performed to evaluate the goodness of the models' results in the forecasting time interval $[T_e, T_e + \Delta T]$ by comparison of the simulated results and observed seismicity. The number of the forecasted earthquakes was evaluated by the N-test and the results are shown in Figures 10 and A9 for both cases. In both cases, for the Anchorage sequence, a more accurate forecast for the number of earthquakes was accomplished by the ETAS model, while for two other sequences the OU law performed better. It should be noted as a result of the branching nature of the ETAS model, it shows a wider confidence interval range compared to the OU law. The obtained results from the M-test demonstrate higher consistency in generating the distribution of the magnitudes for the ETAS model compared to the OU law for both cases, Figures 11 and A11. For the model comparison, the R-test was performed (Figure 12). The α quantile score is higher than the thresholds representing the rejection of the OU hypothesis in favor of the ETAS model. The T-test results are given in Figure 13. The ETAS model performed better in case of the Denali and Kodiak sequences, however, the OU model was more accurate in estimating the rate and the corresponding forecasting performance in case of the Anchorage sequence in its early days. This is also evident when plotting the information gain both for the fixed forecasting time interval $\Delta T = 7$ days with varying training time intervals and in case of the fixed training time interval with varying forecasting time intervals (Figure 13c). The posterior predictive p-value test was performed to assess the fit of the posterior distribution of Bayesian models by comparison of the posterior predictive distribution and the observed data. In Figure 14 the results of Bayesian p-value analysis are given. They indicate that the forecasts based both on the ETAS model and OU formula are consistent in reproducing the maximum event during each corresponding forecasting time interval.

The obtained results indicate that for the sequences analyzed the forecasting based on the ETAS model and the OU formula produce comparable results for shorter time intervals after the mainshocks. However, the EAST model is more realistic in terms of reproducing the seismicity on longer time scales. Moreover, the ETAS model performs better when the mainshock sequence is preceded by a well defined foreshock sequence [3].

The ETAS model typically performs better with increased number of events in the sequence. However, this is limited by the current earthquake catalogues which typically have relatively high level of completeness that results in fewer events.

Author Contributions: Conceptualization, M.S. and R.S.; methodology, M.S. and R.S.; software, M.S. and R.S.; validation, M.S. and R.S.; writing—original draft preparation, M.S.; writing—review and editing, R.S.; visualization, M.S. and R.S.; supervision, R.S.; project administration, R.S.; funding acquisition, R.S. All authors have read and agreed to the published version of the manuscript.

Funding: M.S. and R.S. would like to acknowledge the support from the UWO IDI grant. R.S. was partially supported by NSERC Discovery grant.

Institutional Review Board Statement: Not applicable.

Informed Consent Statement: Not applicable.

Data Availability Statement: The United States Geological Survey (USGS) search engine was used to extract the earthquake catalogue https://earthquake.usgs.gov/earthquakes/search/ (accessed on 18 December 2021). The United States Geological Survey (USGS), was used to obtain the Quaternary fault and fold database for the United States [33]. The mainshock focal mechanism was obtained from the USGS Moment Tensor catalogue [34–36]. The data analysis was performed using a computer code written in Matlab and can be requested from the author.

Acknowledgments: The authors would like to thank Matteo Taroni and two other anonymous reviewers for their useful and constructive comments that helped to improve the paper.

Conflicts of Interest: The authors declare no conflict of interest.

Abbreviations

The following abbreviations are used in this manuscript:

BPD	Bayesian Predictive Distribution
CSEP	Collaboratory for the Study of Earthquake Predictability
ETAS	Epidemic Type Aftershock Sequence
EVD	Extreme Value Distribution
GFT	Goodness of Fit Test
GR	Gutenberg–Richter
MCMC	Markov Chain Monte Carlo

MLE   Maximum Likelihood Estimation
OU   Omori–Utsu
USGS   United States Geological Survey

Appendix A

Table A1. Forecast model parameters distribution from 10,000 MCMC sampling for 2018, Anchorage sequence.

Model	Parameter	Mean	Std	95% CI
Omori–Utsu law, Equation (5),	β	2.077	0.093	$[1.897, 2.261]$
with Exponential frequency	K	144.3	2.838	$[138.8, 149.7]$
magnitude distribution,	c	0.220	0.019	$[0.185, 0.261]$
Equation (3),	p	1.4	0.046	$[1.311, 1.491]$
ETAS model, Equation (6),	β	2.28	0.11	$[2.07, 2.50]$
with Exponential frequency	μ	0.001	0.0003	$[0.0004, 0.0017]$
magnitude distribution,	A	0.104	0.03	$[0.06, 0.16]$
Equation (3),	c	0.028	0.008	$[0.013, 0.045]$
	p	1.113	0.034	$[1.048, 1.184]$
	α	2.42	0.097	$[2.257, 2.659]$

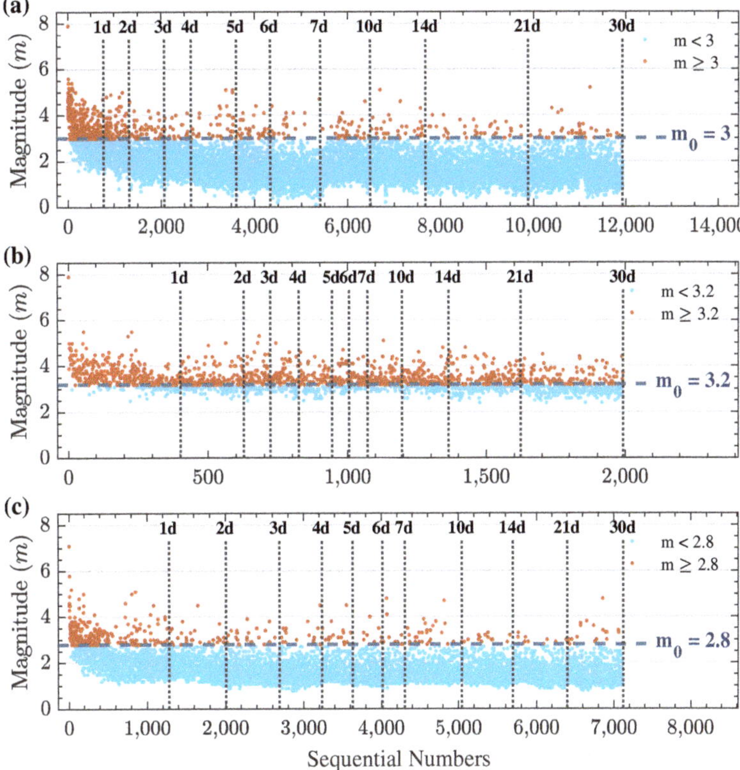

Figure A1. Plot of the magnitudes versus sequential numbers of the earthquakes in the study regions for the three sequences: (**a**) 3 November 2002, Mw 7.9 Denali (**b**) 23 January 2018, Mw 7.9 Kodiak (**c**) 30 November 2018, Mw 7.1 Anchorage. The corresponding times in days after each mainshock are depicted by doted vertical lines.

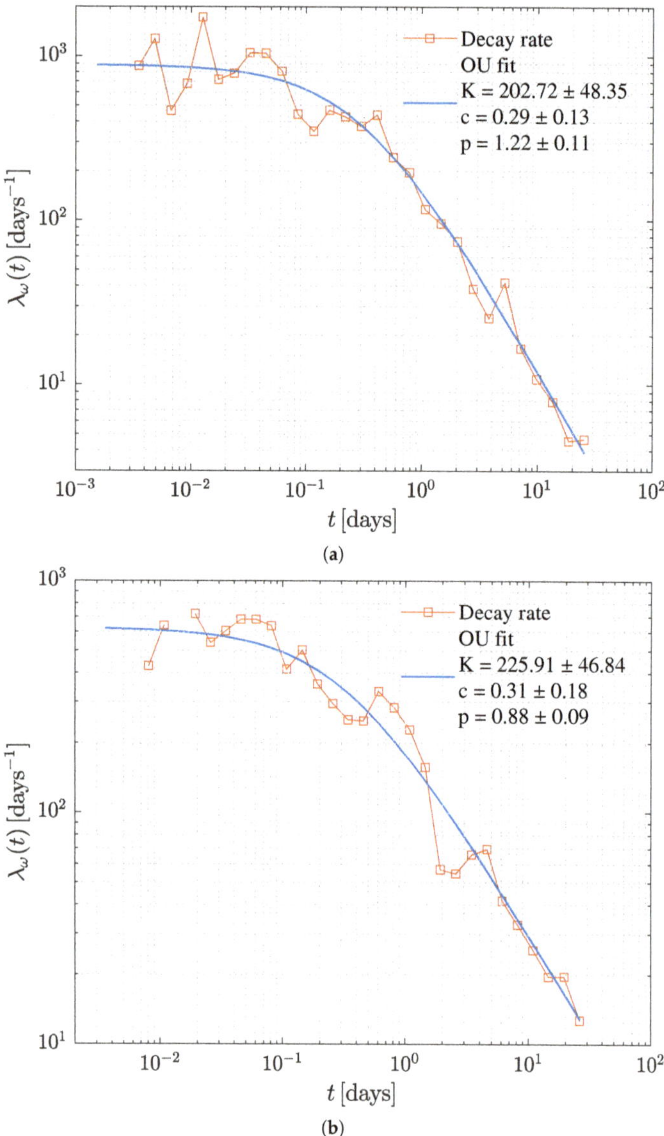

Figure A2. The log-log plot of the earthquake decay rates for: (**a**) the 2002, Mw 7.9 Denali sequence with $m \geq 3.0$; (**b**) the 2018, Mw 7.9 Kodiak sequence with $m \geq 3.2$ are presented as open squares. The blue solid lines are the corresponding fit of the OU law, Equation (5), to the aftershock sequences. The obtained parameters from the OU law, Equation (5), with the 95% confidence intervals are reported in the legends.

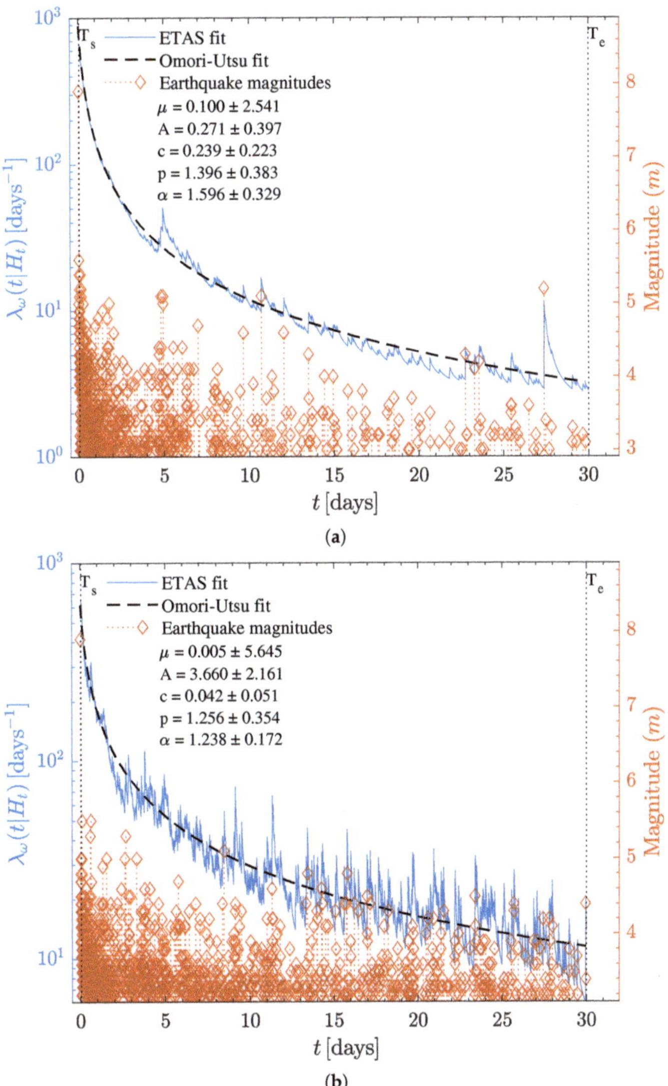

Figure A3. The aftershock sequence and corresponding earthquake magnitude for: (**a**) the 2002, Mw 7.9 Denali sequence with $m \geq 3.0$; (**b**) the 2018, Mw 7.9 Kodiak sequence with $m \geq 3.2$. The ETAS model fit, Equation (6), for the target time interval of $[T_s, T_e] = [0.06, 30]$ is plotted as a solid blue line, and the obtained set of parameters are reported with 95% confidence intervals. The OU law fit, Equation (5), is plotted as a black dashed line for comparison.

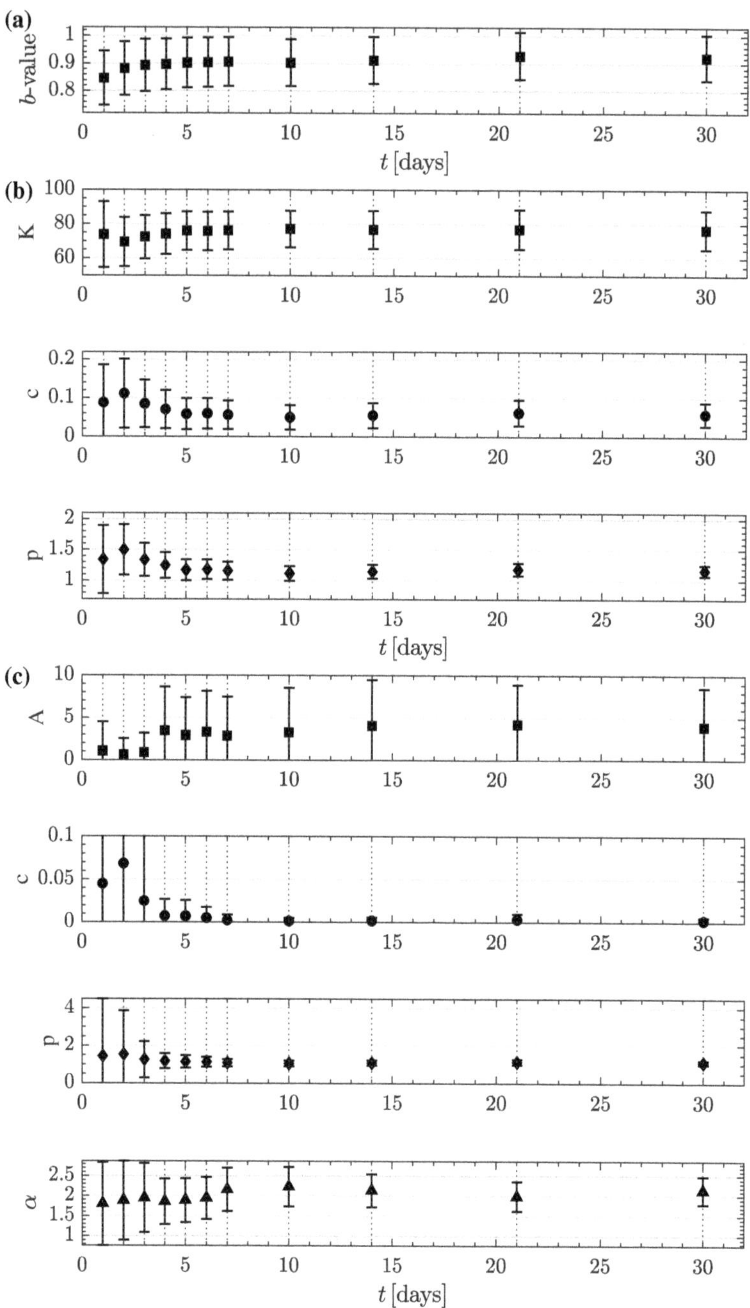

Figure A4. The model parameter estimation during the aftershock sequence of the 2018, Anchorage for all the events with magnitude 2.8 and greater. (**a**) The estimated b-value, Equation (1), (**b**) the parameters $\{K, c, p\}$ of the OU law, Equation (5), and (**c**) the parameters $\{A, c, p, \alpha\}$ of the ETAS model, Equation (6). The error bars represent the 95% confidence intervals during the target time intervals, $\{1, 2, 3, 4, 5, 6, 7, 10, 14, 21, 30\}$, days after the mainshock.

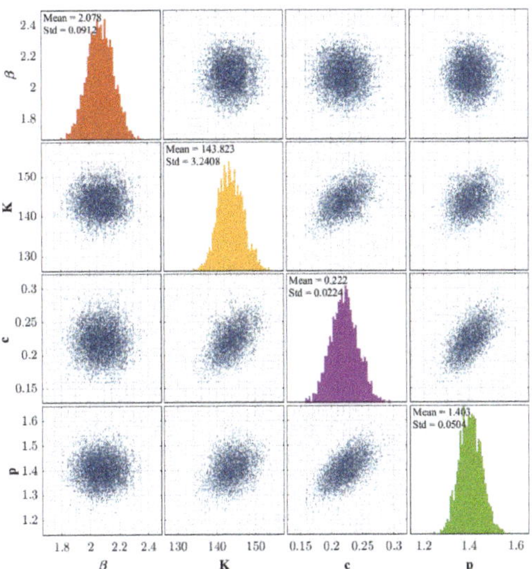

Figure A5. The distribution of each parameter and the matrix plot of the pairs of the OU parameters that computed using the MCMC sampling for the 2018, Mw 7.1 Anchorage sequence with $m \geq 2.8$.

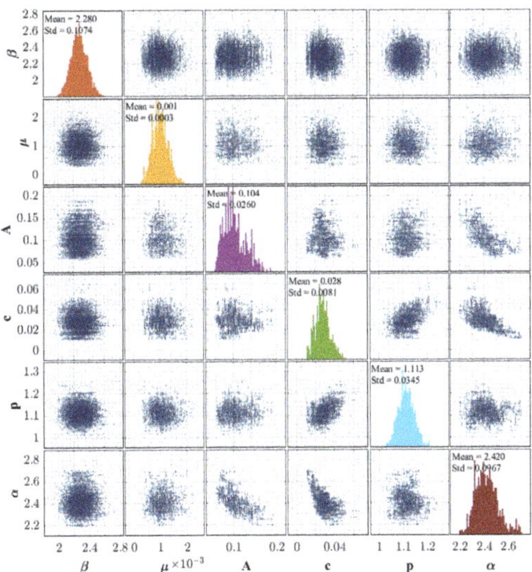

Figure A6. The distribution of each parameter and the matrix plot of the pairs of the ETAS parameters computed using the MCMC sampling for the 2018, Mw 7.1 Anchorage sequence with $m \geq 2.8$.

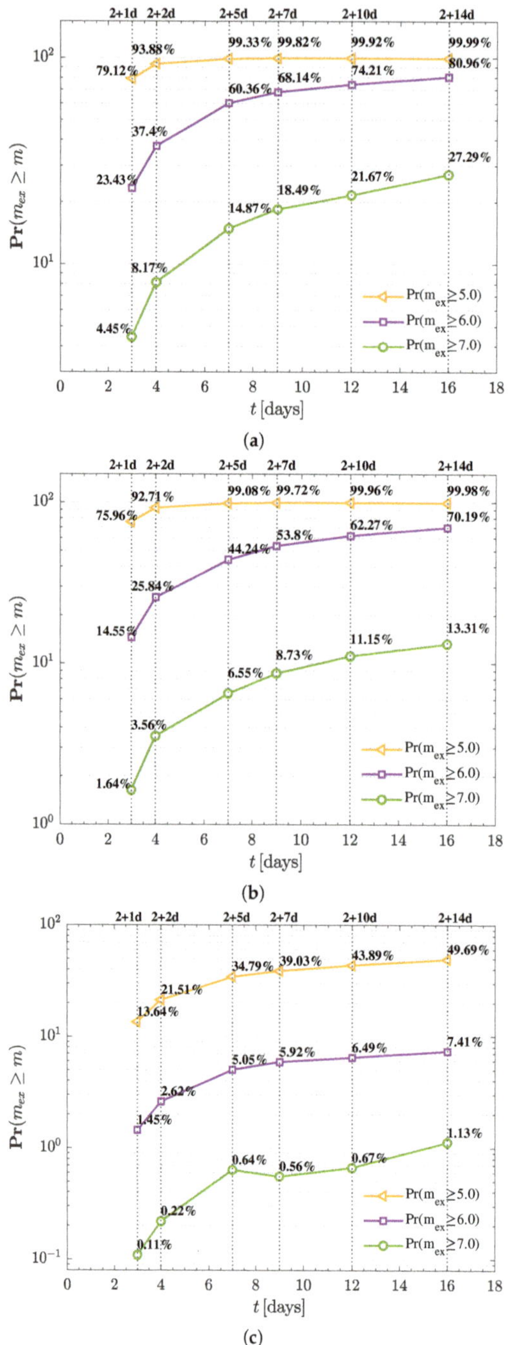

Figure A7. The probabilities to have the largest expected aftershocks to be larger than $m_{ex} \geq 5.0$, 6.0, 7.0 using the BPD, Equation (12), during a constant target time interval of $[T_s, T_e] = [0.06, 2]$ days and for the varying target time intervals. (**a**) The 3 November 2002, Mw 7.9 Denali sequence with $m \geq 3.0$. (**b**) The 23 January 2018, Mw 7.9 Kodiak sequence with $m \geq 3.2$. (**c**) The 30 November 2018, Mw 7.1 Anchorage sequence with $m \geq 2.8$.

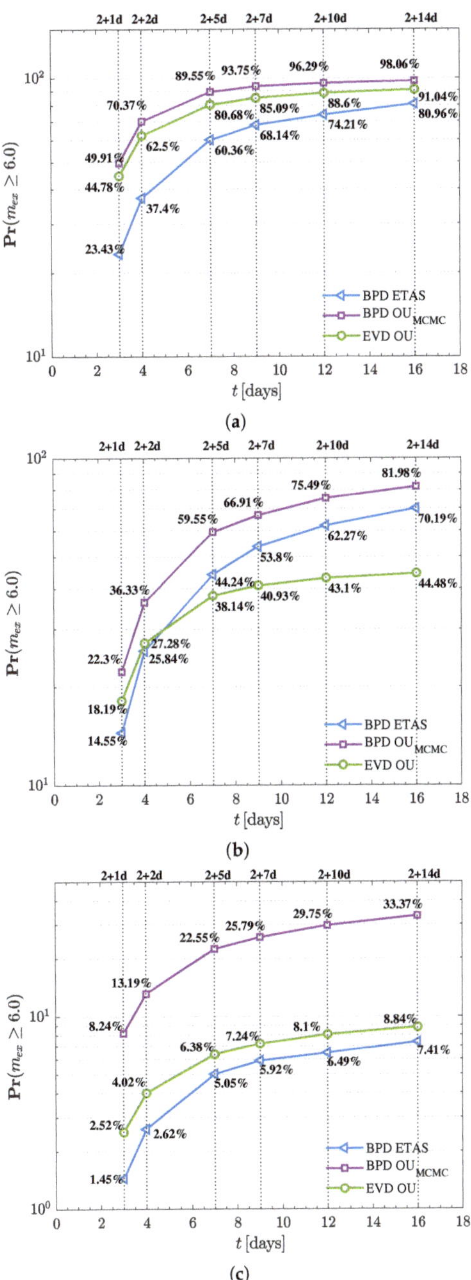

Figure A8. The comparison of the probabilities to have the largest expected aftershock during the target time interval of $[T_s, T_e] = [0.06, 2]$ days for the three sequences: (**a**) 3 November 2002, Mw 7.9 Denali (**b**) 23 January 2018, Mw 7.9 Kodiak (**c**) 30 November 2018, Mw 7.1 Anchorage. The blue triangles are computed using the BPD, Equation (12), with an earthquake decay rate given by the ETAS model, Equation (6). The purple squares are computed using BPD, Equation (12), with an earthquake decay rate given by OU law, Equation (5). The green circles give probabilities computed using the EVD, Equation (11). The aftershock magnitudes are modelled using Equation (3).

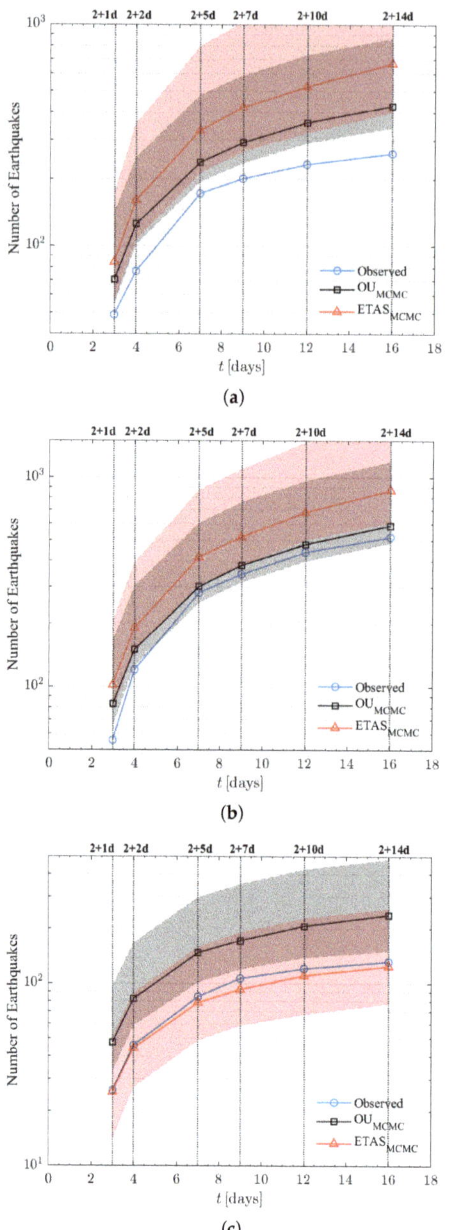

Figure A9. The number of forecasted and observed aftershocks in the forecasting time interval $\Delta T = [1, 2, 5, 7, 10, 14]$ days for (**a**) The 3 November 2002, Mw 7.9 Denali for $m \geq 3.0$; (**b**) The 23 January 2018, Mw 7.9 Kodiak for $m \geq 3.2$; (**c**) The 30 November 2018, Mw 7.1 Anchorage for $m \geq 2.8$ by using the constant target time interval $[T_s, T_e] = [0.06, 2]$ days. The red squares show the average number of forecasted earthquakes using the ETAS model, Equation (6), and the black triangles illustrate the average number of forecasted earthquakes using OU law, Equation (5). The shading bands represent 95% confidence intervals. The blue circles represents the observed number of earthquakes in the forecast time interval.

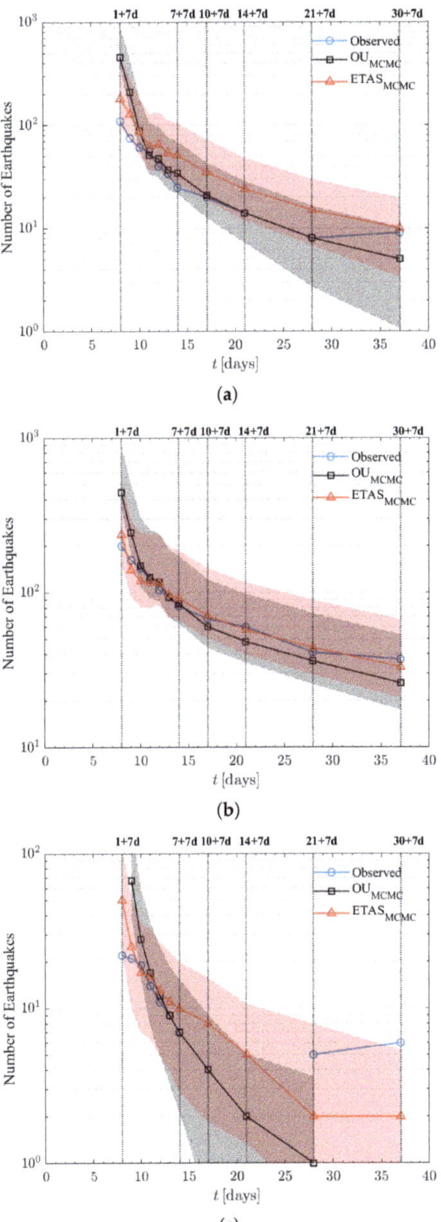

Figure A10. The number of forecasted and observed aftershocks in the forecasting time interval $\Delta T = 7$ days for the three sequences: (**a**) the 3 November 2002, Mw 7.9 Denali sequence for $m \geq 3.5$; (**b**) the 23 January 2018, Mw 7.9 Kodiak sequence for $m \geq 3.5$; (**c**) the 30 November 2018, Mw 7.1 Anchorage sequence for $m \geq 3.5$. The red triangles show the average number of forecasted earthquakes using the ETAS model and the black squares illustrate the average number of forecasted earthquakes using OU law. The shading bands represent 95% confidence intervals. The blue circles represent the observed number of earthquakes in each forecasting time interval.

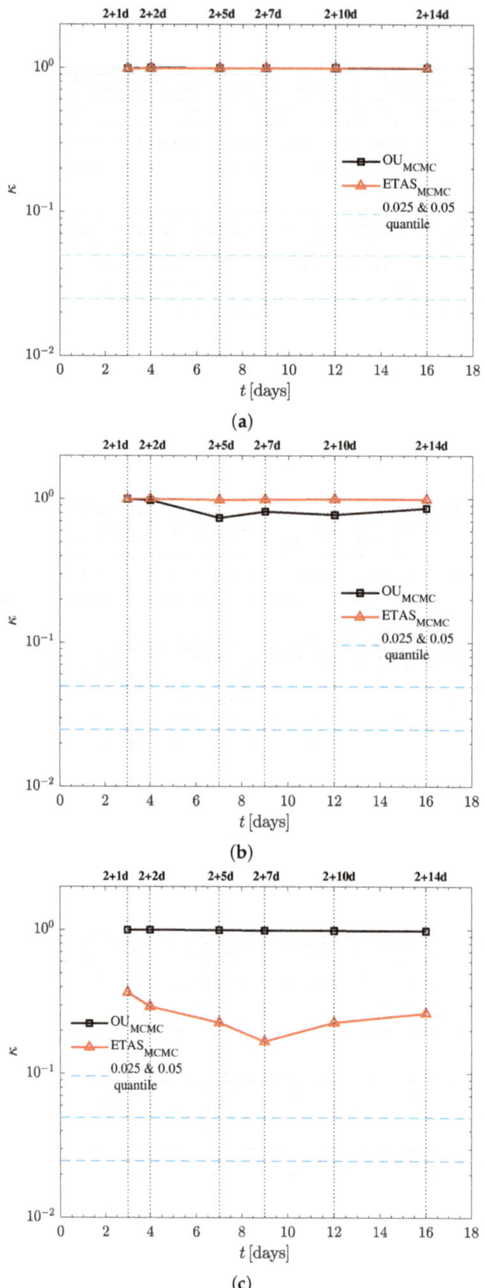

Figure A11. The obtained quantile scores from the M-test for the constant target time interval $[T_s, T_e] = [0.06, 2]$ days for (**a**) The 3 November 2002, Mw 7.9 Denali for $m \geq 3.0$; (**b**) The 23 January 2018, Mw 7.9 Kodiak for $m \geq 3.2$; (**c**) The 30 November 2018, Mw 7.1 Anchorage for $m \geq 2.8$. The red triangles demonstrate the obtained quantile scores from the ETAS model, Equation (6), and the black squares illustrate the quantile scores of OU law, Equation (5). The blue dashed lines represent the 0.025th and 0.05th quantiles.

References

1. Goda, K.; Pomonis, A.; Chian, S.; Offord, M.; Saito, K.; Sammonds, P.; Fraser, S.; Raby, A.; Macabuag, J. Ground motion characteristics and shaking damage of the 11th March 2011 Mw9.0 Great East Japan earthquake. *Bull. Earthq. Eng.* **2013**, *11*, 141–170. [CrossRef]
2. Daniell, J.; Schaefer, A.; Wenzel, F. Losses associated with secondary effects in earthquakes. *Front. Built Environ.* **2017**, *3*, 30. [CrossRef]
3. Shcherbakov, R.; Zhuang, J.; Zöller, G.; Ogata, Y. Forecasting the magnitude of the largest expected earthquake. *Nat. Commun.* **2019**, *10*, 4051. [CrossRef] [PubMed]
4. Shcherbakov, R. Statistics and Forecasting of Aftershocks During the 2019 Ridgecrest, California, Earthquake Sequence. *J. Geophys. Res. Solid Earth* **2021**, *126*, e2020JB020887. [CrossRef]
5. Båth, M. Lateral inhomogeneities of the upper mantle. *Tectonophysics* **1965**, *2*, 483–514. [CrossRef]
6. Vere-Jones, D. A note on the statistical interpretation of Båth's law. *Bull. Seismol. Soc. Am.* **1969**, *59*, 1535–1541. [CrossRef]
7. Vere-Jones, D. Stochastic Models for Earthquake Sequences. *Geophys. J. Int.* **1975**, *42*, 811–826. [CrossRef]
8. Reasenberg, P.A.; Jones, L.M. Earthquake Hazard after a Mainshock in California. *Science* **1989**, *243*, 1173–1176. [CrossRef]
9. Michael, A.J.; McBride, S.K.; Hardebeck, J.L.; Barall, M.; Martinez, E.; Page, M.T.; van der Elst, N.; Field, E.H.; Milner, K.R.; Wein, A.M. Statistical Seismology and Communication of the USGS Operational Aftershock Forecasts for the 30 November 2018 Mw 7.1 Anchorage, Alaska, Earthquake. *Seismol. Res. Lett.* **2019**, *91*, 153–173. [CrossRef]
10. Mignan, A.; Woessner, J. Estimating the magnitude of completeness for earthquake catalogs. *Community Online Resour. Stat. Seism. Anal.* **2012**. [CrossRef]
11. Kagan, Y.Y. Short-term properties of earthquake catalogs and models of earthquake source. *Bull. Seismol. Soc. Am.* **2004**, *94*, 1207–1228. [CrossRef]
12. Peng, Z.G.; Vidale, J.E.; Houston, H. Anomalous early aftershock decay rate of the 2004 Mw 6.0 Parkfield, California, earthquake. *Geophys. Res. Lett.* **2006**, *33*, L17307. [CrossRef]
13. Omi, T.; Ogata, Y.; Hirata, Y.; Aihara, K. Estimating the ETAS model from an early aftershock sequence. *Geophys. Res. Lett.* **2014**, *41*, 850–857. [CrossRef]
14. Page, M.T.; van der Elst, N.; Hardebeck, J.; Felzer, K.; Michael, A.J. Three Ingredients for Improved Global Aftershock Forecasts: Tectonic Region, Time-Dependent Catalog Incompleteness, and Intersequence Variability. *Bull. Seismol. Soc. Am.* **2016**, *106*, 2290–2301. [CrossRef]
15. García, J.; Gómez Rico, M.; Molina Collado, A. A destination-branding model: An empirical analysis based on stakeholders. *Tour. Manag.* **2012**, *33*, 646–661. [CrossRef]
16. Gerstenberger, M.C.; Marzocchi, W.; Allen, T.; Pagani, M.; Adams, J.; Danciu, L.; Field, E.H.; Fujiwara, H.; Luco, N.; Ma, K.F.; et al. Probabilistic Seismic Hazard Analysis at Regional and National Scale: State of the Art and Future Challenges. *Rev. Geophys.* **2020**, *58*, e2019RG000653. [CrossRef]
17. Schorlemmer, D.; Gerstenberger, M.C.; Wiemer, S.; Jackson, D.D.; Rhoades, D.A. Earthquake likelihood model testing. *Seismol. Res. Lett.* **2007**, *78*, 17–29. [CrossRef]
18. Schorlemmer, D.; Werner, M.J.; Marzocchi, W.; Jordan, T.H.; Ogata, Y.; Jackson, D.D.; Mak, S.; Rhoades, D.A.; Gerstenberger, M.C.; Hirata, N.; et al. The Collaboratory for the Study of Earthquake Predictability: Achievements and Priorities. *Seismol. Res. Lett.* **2018**, *89*, 1305–1313. [CrossRef]
19. Zechar, J.D.; Gerstenberger, M.C.; Rhoades, D.A. Likelihood-Based Tests for Evaluating Space-Rate-Magnitude Earthquake Forecasts. *Bull. Seismol. Soc. Am.* **2010**, *100*, 1184–1195. [CrossRef]
20. Taroni, M.; Marzocchi, W.; Schorlemmer, D.; Werner, M.J.; Wiemer, S.; Zechar, J.D.; Heiniger, L.; Euchner, F. Prospective CSEP Evaluation of 1-Day, 3-Month, and 5-Yr Earthquake Forecasts for Italy. *Seismol. Res. Lett.* **2018**, *89*, 1251–1261. [CrossRef]
21. Nanjo, K.Z.; Tsuruoka, H.; Yokoi, S.; Ogata, Y.; Falcone, G.; Hirata, N.; Ishigaki, Y.; Jordan, T.H.; Kasahara, K.; Obara, K.; et al. Predictability study on the aftershock sequence following the 2011 Tohoku-Oki, Japan, earthquake: First results. *Geophys. J. Int.* **2012**, *191*, 653–658. [CrossRef]
22. Cattania, C.; Werner, M.J.; Marzocchi, W.; Hainzl, S.; Rhoades, D.; Gerstenberger, M.; Liukis, M.; Savran, W.; Christophersen, A.; Helmstetter, A.; et al. The Forecasting Skill of Physics-Based Seismicity Models during the 2010–2012 Canterbury, New Zealand, Earthquake Sequence. *Seismol. Res. Lett.* **2018**, *89*, 1238–1250. [CrossRef]
23. Rhoades, D.A.; Liukis, M.; Christophersen, A.; Gerstenberger, M.C. Retrospective tests of hybrid operational earthquake forecasting models for Canterbury. *Geophys. J. Int.* **2016**, *204*, 440–456. [CrossRef]
24. Rhoades, D.A.; Christophersen, A.; Gerstenberger, M.C.; Liukis, M.; Silva, F.; Marzocchi, W.; Werner, M.J.; Jordan, T.H. Highlights from the First Ten Years of the New Zealand Earthquake Forecast Testing Center. *Seismol. Res. Lett.* **2018**, *89*, 1229–1237. [CrossRef]
25. Lay, T.; Ye, L.; Bai, Y.; Cheung, K.F.; Kanamori, H. The 2018 MW 7.9 Gulf of Alaska earthquake: Multiple fault rupture in the Pacific plate. *Geophys. Res. Lett.* **2018**, *45*, 9542–9551. [CrossRef]
26. Ruppert, N.A.; Rollins, C.; Zhang, A.; Meng, L.; Holtkamp, S.G.; West, M.E.; Freymueller, J.T. Complex faulting and triggered rupture during the 2018 MW 7.9 offshore Kodiak, Alaska, earthquake. *Geophys. Res. Lett.* **2018**, *45*, 7533–7541. [CrossRef]
27. West, M.E.; Bender, A.; Gardine, M.; Gardine, L.; Gately, K.; Haeussler, P.; Hassan, W.; Meyer, F.; Richards, C.; Ruppert, N.; et al. The 30 November 2018 M w 7.1 Anchorage earthquake. *Seismol. Res. Lett.* **2020**, *91*, 66–84. [CrossRef]

28. Gomberg, J. Slow-slip phenomena in Cascadia from 2007 and beyond: A review. *Geol. Soc. Am. Bull.* **2010**, *122*, 963–978. [CrossRef]

29. Bhattacharya, P.; Phan, M.; Shcherbakov, R. Statistical Analysis of the 2002 Mw 7.9 Denali Earthquake. *Bull. Seismol. Soc. Am.* **2011**, *101*, 2662–2674. [CrossRef]

30. Vere-Jones, D. Foundations of Statistical Seismology. *Pure Appl. Geophys.* **2010**, *167*, 645–653. [CrossRef]

31. Utsu, T. A statistical study on the occurrence of aftershocks. *Geophys. Mag.* **1961**, *30*, 521–605.

32. Ogata, Y. Statistics of Earthquake Activity: Models and Methods for Earthquake Predictability Studies. *Annu. Rev. Earth Planet. Sci.* **2017**, *45*, 497–527. [CrossRef]

33. USGS. Quaternary Fault and Fold Database. 2006. Available online: https://www.usgs.gov/natural-hazards/earthquake-hazards/faults/ (accessed on 18 December 2021).

34. USGS. Mw 7.9 Central Alaska. 2015. Available online: https://earthquake.usgs.gov/earthquakes/eventpage/ak002e435qpj/executive (accessed on 18 December 2021).

35. USGS. Mw 7.9 Kodiak Alaska. 2018. Available online: https://earthquake.usgs.gov/earthquakes/eventpage/us2000cmy3/executive (accessed on 18 December 2021).

36. USGS. Mw 7.1 Anchorage Alaska. 2018. Available online: https://earthquake.usgs.gov/earthquakes/eventpage/ak20419010/executive (accessed on 18 December 2021).

37. Krabbenhoeft, A.; von Huene, R.; Miller, J.J.; Lange, D.; Vera, F. Strike-slip 23 January 2018 MW 7.9 Gulf of Alaska rare intraplate earthquake: Complex rupture of a fracture zone system. *Sci. Rep.* **2018**, *8*, 13706. [CrossRef]

38. Guo, Y.; Miyakoshi, K.; Tsurugi, M. Simultaneous rupture on conjugate faults during the 2018 Anchorage, Alaska, intraslab earthquake (MW 7.1) inverted from strong-motion waveforms. *Earth Planets Space* **2020**, *72*, 176. [CrossRef]

39. Liu, C.; Lay, T.; Xie, Z.; Xiong, X. Intraslab deformation in the 30 November 2018 Anchorage, Alaska, MW 7.1 earthquake. *Geophys. Res. Lett.* **2019**, *46*, 2449–2457. [CrossRef]

40. Gutenberg, B.; Richter, C.F. Frequency of earthquakes in California. *Bull. Seismol. Soc. Am.* **1944**, *4*, 185–188. [CrossRef]

41. Bender, B. Maximum-likelihood estimation of *b*-values for magnitude grouped data. *Bull. Seismol. Soc. Am.* **1983**, *73*, 831–851. [CrossRef]

42. Tinti, S.; Mulargia, F. Confidence intervals of *b*-values for grouped magnitudes. *Bull. Seismol. Soc. Am.* **1987**, *77*, 2125–2134.

43. Omori, F. On after-shocks of earthquakes. *J. Coll. Sci. Imp. Univ. Tokyo* **1894**, *7*, 113–200.

44. Ogata, Y. Estimation of the Parameters in the Modified Omori Formula for Aftershock Frequencies by the Maximum-Likelihood Procedure. *J. Phys. Earth* **1983**, *31*, 115–124. [CrossRef]

45. Ogata, Y. Seismicity analysis through point-process modeling: A review. *Pure Appl. Geophys.* **1999**, *155*, 471–507. [CrossRef]

46. Ogata, Y. Statistical-Models For Earthquake Occurrences And Residual Analysis For Point-Processes. *J. Am. Stat. Assoc.* **1988**, *83*, 9–27. [CrossRef]

47. Daley, D.J.; Vere-Jones, D. *An Introduction to the Theory of Point Processes*, 2nd ed.; Springer: New York, NY, USA, 2003; Volume 1.

48. Shcherbakov, R.; Zhuang, J.; Ogata, Y. Constraining the magnitude of the largest event in a foreshock-mainshock-aftershock sequence. [CrossRef]

49. Rhoades, D.A.; Schorlemmer, D.; Gerstenberger, M.C.; Christophersen, A.; Zechar, J.D.; Imoto, M. Efficient testing of earthquake forecasting models. *Acta Geophys.* **2011**, *59*, 728–747. [CrossRef]

50. Wiemer, S.; Wyss, M. Minimum magnitude of completeness in earthquake catalogs: Examples from Alaska, the western United States, and Japan. *Bull. Seismol. Soc. Am.* **2000**, *90*, 859–869. [CrossRef]

51. Zhuang, J.; Ogata, Y.; Wang, T. Data completeness of the Kumamoto earthquake sequence in the JMA catalog and its influence on the estimation of the ETAS parameters. *Earth Planets Space* **2017**, *69*, 1–12. [CrossRef]

52. Gulia, L.; Rinaldi, A.P.; Tormann, T.; Vannucci, G.; Enescu, B.; Wiemer, S. The effect of a mainshock on the size distribution of the aftershocks. *Geophys. Res. Lett.* **2018**, *45*, 13–277. [CrossRef]

53. Spassiani, I.; Marzocchi, W. An Energy-Dependent Earthquake Moment–Frequency Distribution. *Bull. Seismol. Soc. Am.* **2021**, *111*, 762–774. [CrossRef]

Article

Identification and Temporal Characteristics of Earthquake Clusters in Selected Areas in Greece

Polyzois Bountzis [1,*], Eleftheria Papadimitriou [1] and George Tsaklidis [2]

[1] Geophysics Department, Aristotle University of Thessaloniki, GR 541 24 Thessaloniki, Greece; ritsa@geo.auth.gr

[2] Statistics and Operational Research Department, Aristotle University of Thessaloniki, GR 541 24 Thessaloniki, Greece; tsaklidi@math.auth.gr

[*] Correspondence: pmpountzp@geo.auth.gr

Abstract: The efficiency of earthquake clustering investigation is improved as we gain access to larger datasets due to the increase of earthquake detectability. We aim to demonstrate the robustness of a new clustering method, MAP-DBSCAN, and to present a comprehensive analysis of the clustering properties in three major seismic zones of Greece during 2012–2019. A time-dependent stochastic point model, the Markovian Arrival Process (MAP), is implemented for the detection of change-points in the seismicity rate and subsequently, a density-based clustering algorithm, DBSCAN, is used for grouping the events into spatiotemporal clusters. The two-step clustering procedure, MAP-DBSCAN, is compared with other existing methods (Gardner-Knopoff, Reasenberg, Nearest-Neighbor) on a simulated earthquake catalog and is proven highly competitive as in most cases outperforms the tested algorithms. Next, the earthquake clusters in the three areas are detected and the regional variability of their productivity rates is investigated based on the generic estimates of the Epidemic Type Aftershock Sequence (ETAS) model. The seismicity in the seismic zone of Corinth Gulf is characterized by low aftershock productivity and high background rates, indicating the dominance of swarm activity, whereas in Central Ionian Islands seismic zone where main shock-aftershock sequences dominate, the aftershock productivity rates are higher. The productivity in the seismic zone of North Aegean Sea vary significantly among clusters probably due to the co-existence of swarm activity and aftershock sequences. We believe that incorporating regional variations of the productivity into forecasting models, such as the ETAS model, it might improve operational earthquake forecasting.

Keywords: seismicity clustering; DBSCAN algorithm; markovian arrival processes; statistical seismology

Citation: Bountzis, P.; Papadimitriou, E.; Tsaklidis, G. Identification and Temporal Characteristics of Earthquake Clusters in Selected Areas in Greece. *Appl. Sci.* **2022**, *12*, 1908. https://doi.org/10.3390/app12041908

Academic Editor: Amadeo Benavent-Climent

Received: 31 December 2021
Accepted: 1 February 2022
Published: 11 February 2022

Publisher's Note: MDPI stays neutral with regard to jurisdictional claims in published maps and institutional affiliations.

1. Introduction

Earthquake clustering is an essential property of seismicity and is manifested as the concentration of earthquakes in space and time. Due to the improvement of seismic monitoring worldwide and the development of new powerful algorithms for earthquake detectability [1] additional information is available, which is crucial for reliable regional estimates of aftershock forecasting probabilities [2,3] and the determination of faulting geometry [4,5], among others. In addition to the necessity of cluster identification, for many studies it is important to reliably separate the background seismicity from clustered events for the development of long-term seismic hazard maps [6–8] or the regional optimization of background rates [9].

Among the methods that are available for the detection of seismic clusters, one of the most widely used is the window-based approach [10] with known drawback the large gaps after the occurrence of strong earthquakes [11]. The link-based model [12] considers stress redistribution and the Omori law for the determination of the spatiotemporal interactions among earthquakes. The stochastic declustering method of Zhuang et al. [13] is based on the modeling of earthquake occurrences by the ETAS model [14,15], where events are separated into background and clustered ones according to the estimated probabilities. Important clustering features can be inferred using the stochastic algorithm [16,17], whereas it can be also used for declustering to optimize the background seismicity rate estimates [9,18]. Marsan et al. [19] introduced an ETAS model with a time-dependent background component for the detection of aseismic transients. The modified ETAS model is used efficiently to reveal both main shock-aftershocks and earthquake swarms [20,21]. Finally, the Nearest-Neighbor metric proposed by Baiesi and Paczuski [22] adopts a non-parametric definition of a cluster considering the space-time-magnitude proximity among earthquakes. Zaliapin and Ben-Zion [23] introduced a binary threshold, η, according to which the earthquakes are classified into clusters and background seismicity. They applied the algorithm on a global scale, revealing a link between the clustering properties of seismicity and the heat flow level [24]. The method is proven to be efficient in detecting main shock–aftershock sequences in Northeastern Italy [25], as well as earthquake repeaters in the Sea of Marmaras [26]. Bayliss et al. [27] introduced a new approach to create probabilistic cluster networks based on the intersection between the background and clustered component of the nearest-neighbor distances.

Another approach is based on the assumption of a common physical trigger during a seismic sequence, expressed by fluctuations in the occurrence rate [28,29]. In our method, change-points of the intensity rate in the earthquake occurrences are detected with the use of the MAP model [30]. The temporal distribution of the events is approximated essentially by a non-homogeneous Poisson process, N_t, with a piece-wise constant intensity rate determined by the underlying Markov process, J_t. Simulation studies and applications on real datasets showed that the model efficiently identifies the changes in the earthquake occurrence rates [31]. Recent works by Lu [32] and Benali et al. [33] are based on non-stationary Poisson models whose rate is modulated by a hidden Markov process to determine a set of change-points for seismicity rate. Concerning earthquake clustering, Bountzis et al. [34] proposed the combination of the MAP model with a density-based clustering algorithm, DBSCAN [35], for the detection of earthquake clusters in the spatiotemporal domain. They used the method on a micro-seismicity earthquake catalog in central Ionian Islands, Greece, efficiently revealing the clustered seismicity.

Several studies suggest that the clustering properties of seismicity (spatiotemporal distribution, productivity rates) might be controlled by the tectonic regime. Llenos and Michael [36] showed that the adoption of region-specific aftershock parameters can improve forecast estimates, as the information from the tectonic region is particularly useful, and suggest the determination of clustering features in smaller regions where high-quality earthquake data are available. More recently, Hardebeck et al. [37] updated the generic parameters of sequences in California incorporating the regionalization of the former work for their determination. In this way, there was an improvement of the aftershock forecasts' accuracy. The temporal ETAS model assumes that seismicity is evolving in the form of independent events (background seismicity) who generate their aftershocks with each one producing their own. It incorporates two empirical laws, the Omori–Utsu law [38] and the productivity law used to explain the distribution of aftershocks, as well as a stable Poissonian rate for the background seismicity. Our work utilizes the estimated parameters of the ETAS model to investigate regional variabilities in the productivity of the sequences and gain insights into the involved triggering mechanisms [39–41].

The aim of this study is firstly to demonstrate the efficiency of the proposed clustering method to separate triggered from background seismicity and subsequently to investigate and compare the clustering properties among three major seismic zones of Greece. In particular, we focus on the statistical analysis of the detected clusters based on the ETAS model producing generic and sequence specific parameters for each area.

The paper is organized as follows. In Section 2, the MAP-DBSCAN method for the identification of the clusters is described and simulation results for the evaluation of the method are deployed. MAP is used as a tool for the detection of changes in the seismicity rate and DBSCAN is implemented to reveal spatially high-density areas. In Section 3, the study area along with the datasets used in the application are presented. Finally, in Section 4, details on the detected clusters for each area are given and the ETAS regional parameters based on the identified clusters are derived. The regional variability of the clustering properties among the three areas is investigated based on the generic estimates of ETAS parameters (a, K, p, c and μ) and sequence-specific parameters. A brief discussion of the results is given in Section 5 and the main conclusions are presented in Section 6.

2. MAP-DBSCAN Method

2.1. MAP as a Tool for the Detection of Seismicity Rate Changes

In the first step, the temporal distribution of seismicity is approximated by a stochastic point model, the Markovian Arrival Process. The MAP is a two-dimensional Markov process $(N_t, J_t)_{t \in R+}$, where N_t counts the number of earthquakes up to time t that occur with a rate λ_t. Its value is associated with the unobservable states $i = 1, \ldots, K$, of the Markov process J_t. In particular, when the process J_t is in state i, earthquakes occur according to a Poisson process with rate λ_i and, therefore, the sojourn time in this state follows an exponential distribution with expected value $1/\lambda_i$. When an earthquake occurs, the MAP can transit with probability p_{ij} to another state j, so now, earthquakes occur according to a Poisson process with rate λ_j, or remain in the current state i with probability p_{ii}.

To represent the MAP model, we need the $K \times K$ rate matrices $\mathbf{D}_0$ and $\mathbf{D}_1$, where $\mathbf{D}_0$ is a diagonal matrix whose non-negative elements we denote as $\lambda_1, \ldots, \lambda_K$, and which correspond to the K Poissonian rates, each one assigned to a hidden state of process J_t, and $\mathbf{D}_1$ consists of the transition rates among the states, along with the occurrence of an earthquake, which we denote as q_{ij}. Additional details on the estimation procedure of the MAP model and its properties are given in Appendix A.

Concerning our clustering algorithm, we are interested in the evaluation of the transitions among the hidden states, namely, the detection of changes in the seismicity rate. In particular, we define a rate threshold, λ_{thr}, according to which a potential sequence starts when the rate of the counting process N_t achieves $\lambda_t > \lambda_{thr}$ and ends as soon as the process J_t moves for the first time to a state with a Poisson rate below that threshold. Our main assumption is that each state corresponds to a distinct evolution phase of a seismic sequence, independently of its underlying mechanism. In this way, the model has the ability to approximate the temporal evolution of earthquake catalogs that incorporate both aftershock sequences and earthquake swarms [31], as well as datasets with non-stationary characteristics [34].

2.2. Temporal Constraints

The earthquakes above the defined rate threshold comprise the potential clusters. However, results on methods that are based solely on changes in the seismicity rate can sometimes be misleading. One such case is when the rate at the tail of aftershock sequences has reached the level of the background seismicity, so it becomes difficult to discriminate these events from background ones. One similar case is related to the sparse foreshock activity, which, as it is shown in Lippiello et al. [42], exhibits significantly smaller frequency than the aftershock activity. Therefore, a day rule, dt, is assigned in the sense that events in $\pm dt$ from the potential cluster are included within. Another case that we observed is related to the existence of fluctuations during a seismic excitation, when the seismic activity that is triggered by the same underlying mechanism is divided into smaller clusters. For this reason, we assign a time window, T, so that clusters in temporal distance smaller than or equal to T are merged into one.

2.3. DBSCAN Algorithm

The merged clusters comprise seismicity concentrated in time. However, events with temporal proximity can be spatially sparse and are falsely assigned into the same cluster. To overcome this ambiguity, a density-based clustering algorithm, DBSCAN, is applied to separate events in space based on a distance metric on the earthquakes' epicentral distribution. Depending on the adopted distance metric, the algorithm can be used for grouping events with waveform similarities [4] as well as earthquakes with related rupture styles and orientations (focal mechanism clustering) [43]. Density-based algorithms search for areas where the event density exceeds a threshold, ϵ. The boundaries of these areas are set where the spatial density falls below that threshold. The DBSCAN algorithm in particular requires as input two parameters, the upper threshold, ϵ, and the minimum number of neighboring events, N_{pts}. A cluster is defined if an earthquake i exists along with at least N_{pts} events within distance $d \leq \epsilon$, including itself. Earthquake i is then considered a core point of the cluster and the algorithm moves to the investigation of the other events. If N_{pts} neighbors are identified, they are also considered core events; otherwise, they consist of the boundary points of the cluster and the algorithm stops. Events that have not been assigned to any cluster at the end of the procedure are included to the background seismicity and are merged with events that occurred during periods with estimated rate under the rate threshold, λ_{thr}. In this way, the algorithm can remove events that are sparsely distributed in space. It has been efficiently applied for detecting similarities among earthquake locations, origin times and focal mechanisms [34,44]. An advantage of the algorithm is that it does not require as input a predefined number of earthquake clusters, such as the k-means algorithm, where further optimization techniques for the determination of the clusters number are necessary [45].

2.4. Performance Evaluation

2.4.1. ETAS Framework

The efficiency of the method to correctly identify spatio-temporal correlated seismicity is evaluated on a simulated ETAS catalog, where the underlying structure of the clusters is known a priori. Additionally, we aim to demonstrate the performance of the method compared to widely used clustering algorithms. In particular, our approach is compared with the Nearest-Neighbor (NN), Gardner and Knopoff (GK) window-based and Reasenberg (RB) link-based algorithms. A detailed review on each one of them is given in Appendix B.

The ETAS model belongs to a wide class of branching processes where the occurrence rate of earthquakes, known as intensity function, depends on the history of all previous seismicity and consists of two parts given by

$$\lambda(t, \mathbf{x} = (x,y)/H_t) = \mu(\mathbf{x}) + K \sum_{j:t_j<t,x_j\in\Sigma_0} e^{a(m_j-m_c)} g(t) f(\mathbf{x}), \tag{1}$$

where j runs over all past earthquakes with magnitude larger than or equal to m_c. The intensity function is evaluated at each event that occurred during the time interval $[t_{in}, T]$ and inside the target region $\Sigma \subseteq \Sigma_0$. However, events in the broader region Σ_0 and time interval $[t_0, T]$, with $t_0 < t_{in}$, can be considered triggering events, therefore, they should be included in the evaluation of $\lambda(t, \mathbf{x})$. The first term of the right-hand side of Equation (1) expresses the background seismicity $\mu(\mathbf{x})$ which is assumed stationary in time (mother events) and clustered in space due to the fault network geometry. The latter term represents the space-time-dependent seismicity (daughter events) expressed through the following empirical laws:

- the productivity law, $Ke^{a(m_j-m_c)}$, which gives the number of aftershocks triggered by a main shock with magnitude m_j;
- the modified Omori law, $g(t) = (p-1)c^{(p-1)}(t - t_j + c)^{-p}$, with $p > 1$, which describes the temporal decay of aftershocks;
- the spatial distribution of aftershocks, $f(\mathbf{x}/M) = \frac{q-1}{\pi d(m_j)^{q-1}}[\|\mathbf{x} - x_j\|_2^2 + d(m_j)]^{-q}$ with $q > 1$, and $d(m_j) = d_0 10^{\gamma(m_j-m_c)}$, which assumes an isotropic distribution of aftershocks around the main shock.

Each mother event generates its daughters (first generation), the daughters generate their own descendants (second generation), and so on. In this way, a cluster is defined as a sequence of events with a common mother event (first event in the cluster). There are also single events, i.e., mother events without any subsequent triggered earthquake.

2.4.2. Simulation Procedure

For the simulation of an ETAS earthquake catalog, first, we need to generate the mother events. The heterogeneity in space can be preserved from the spatial coordinates of a declustered earthquake catalog. In particular, we implement a declustering procedure to the earthquake catalog of Greece ($\Sigma_0 = [19°E - 29°E] \times [33.7°N - 42°N]$) for earthquakes with $m_c = 2.5$ during the period of 2011–2019, using the NN method, and then we produce N_{main} mother events according to a Poisson distribution with mean value equal to the number of the identified background events. Their coordinates are sampled with replacement from the declustered catalog by adding a random factor. The occurrence times are simulated from a uniform distribution $U(t_0, T)$, where $t_0 = 0$ and $T = 20$ years.

The magnitudes are independent from the earthquakes' spatial and temporal distribution and follow the Gutenberg–Richter (GR) law truncated from the left at the completeness magnitude, $m_c = 2.5$, and from the right at the maximum observed magnitude of the instrumental earthquake records in Greece plus a small factor, $m_{max} = 7.8$. The functional form of their distribution is the following, $s(m) = (\beta e^{-\beta m})/(e^{-\beta m_c} - e^{-\beta m_{max}})$, where β relates to the b-value of the GR law with $\beta = b\log(10)$, and we chose $b = 1.0$. After the generation of the background events, we simulate their aftershock number, following Poisson distribution with expected rate equal to the productivity of the model, $k(m_j) = Ke^{a(m_j - m_c)}$. Their occurrence times are sampled from the modified Omori law, $g(t)$, and the locations from the isotropic spatial distribution function, $f(\mathbf{x})$. For next-generation daughters, the triggering step is repeated until there are no more generated events. In Table 1, we give the parameter set that produced the ETAS catalog.

Table 1. ETAS parameters used for the synthetic earthquake catalog with $m_c = 2.5$. The target area is Corinth Gulf, Greece, with $\Sigma x[t_{in}, T] = [21.3°E - 23.2°E] \times [37.9°N - 38.6°N] \times [2, 20]$. The number of clustered events is $N = 4253$ and the number of mother events is $N_{bg} = 1595$.

Parameter		Parameter	
K	0.1	d	2.41×10^{-5}
a	2.19	q	1.805
p	1.13	γ	0.59
c	0.024 (days)	μ (events/day)	4.50

2.4.3. Evaluation

The ETAS earthquake catalog is divided into either single events or mother events with their descendants. We define by $\mathbf{X} = \{\mathbf{X}_k\}_{k=1,...,N_c}$ the true partition of the catalog, where N_c corresponds to the number of clusters, and with $\mathbf{Y}_i = \{\mathbf{Y}_n\}_{n=1,...,N_{y_i}}$, $i = 1,...,K$, the partition after the implementation of the MAP-DBSCAN and the other $K-1$ methods, including different tuning of the algorithm's parameters. N_{y_i} is the number of clusters after the implementation of method i.

Next, we define the Jaccard index [46], which is a measure to quantify the overlap between two partitions, in our case, the true one of the ETAS catalog, $\mathbf{X}$, and the one of the i-th implemented algorithm, $\mathbf{Y}_i$. The Jaccard index is expressed by $J_1(\mathbf{X}, \mathbf{Y}) = a_{11}/(a_{11} + a_{10} + a_{01})$, where a_{11} indicates the number of pairs of elements which are correctly assigned into the same cluster (true links), a_{01}—the number of pairs of elements which are in the same cluster in the ETAS catalog and in different clusters in the estimated one (missed links) and a_{10}—the number of pairs of elements which are wrongly identified as clustered events (false links). If all the initial clusters are correctly identified by the implemented method, then $a_{10} = 0 = a_{01}$ and $J_1(\mathbf{X}, \mathbf{Y}) = 1$. Conversely, if all pairs are wrongly identified as clustered or independent, then $a_{11} = 0$ and, as a consequence, $J_1(\mathbf{X}, \mathbf{Y}) = 0$.

In addition, we introduce a generalization of the Jaccard index, $J_2(\mathbf{X}, \mathbf{Y}) = b_{11}/(b_{11} + b_{10} + b_{01})$, to identify the partition $\mathbf{Y}$ with the best discrimination between the background seismicity and clustered elements, following the definition in Lippiello and Bountzis [47]. We consider as background seismicity single events and the mother events of each cluster, i.e., the one that initiated a cascade of events. Here, b_{11} represents the number of common background events in the two partitions, b_{10} is the number of elements wrongly identified as mother events in the partition $\mathbf{Y}$, whereas b_{01} corresponds to the number of true mother events identified as clustered elements in the partition $\mathbf{Y}$. In Table 2, we show the Jaccard index values ($J_i, i = 1, 2$) after the implementation of the different clustering algorithms. In particular, for the MAP-DBSCAN algorithm we show the one with the best results in terms of the Jaccard index (MAP-DBSCAN27). In Appendix B, the results for all input parameters are given.

Table 2. $J_i, i = 1, 2$, values for 3 parameter sets (PS) of RB and GK algorithms, respectively, and the corresponding values of the MAP-DBSCAN and NN methods.

PS	RB1	RB2	RB3	GK1	GK2	GK3	MAP-DBSCAN (PS27)	NN
J_1	0.530	0.593	0.648	0.382	0.397	0.585	0.627	0.756
J_2	0.612	0.630	0.617	0.418	0.192	0.676	0.647	0.727

The window-based method removes all the events within $d(M)$ kilometers and $t(M)$ days after a main shock with magnitude M. We used three different temporal and spatial intervals, given by Equations (A2) (GK1), (A3) (GK2) and (A4) (GK3). The Reasenberg algorithm combines a deterministic spatial window and a probabilistic temporal one, determined by the Omori law. We used three sets of parameters in the ZMAP tool. RB1 (Table A1) corresponds to the original parameters proposed in Reasenberg [12]. In the second set, RB2, we extend the spatial zone by increasing the factor r_{fact} from 10 to 20 km, whereas, in the third set, RB3, we also extend the temporal window modifying the parameters τ_{min} and τ_{max} (see Table A1). The NN method separates seismicity into background and clustered events according to the bimodal distribution of the rescaled time and distance metrics. The algorithm needs as input two parameters, the b-value and the fractal dimension of the earthquake locations, which are considered equal to $b = 1.0$ and $d_f = 1.51$, respectively. For the MAP-DBSCAN method, first, the optimal MAP model and the corresponding rate threshold ($\lambda_{thr} = \lambda_1$) are determined. Then, different temporal constraints are tested for the merging of the potential clusters (consecutive events above the rate threshold) and, subsequently, the DBSCAN is implemented. We set the minimum number of neighbors equal to $N_{pts} = 2$, similar to the minimum size of a cluster that can be given as output from the other algorithms. For the distance threshold, ϵ, we tested a wide range of possible values (see Table A2).

The NN method shows the best performance in the construction of the clusters ($J_1 = 0.756$) and in the detection of the mother events ($J_2 = 0.727$). This is also evident by its cumulative number of background seismicity (purple line in Figure 1a), which is the closest one to the initial catalog (dotted black line). The temporal evolution of background seismicity is shown in Figure 1b–f across the longitude for ease of reading as west–east normal faults dominate the area. For the NN method, no large gaps are evident in the space-time evolution of the declustered seismicity, although there is a significant concentration of events between the 7th and 8th year of the catalog, which is also persistent in both RB2 and GK3 methods (orange ellipses in Figure 1d–f) and less apparent on MAP-DBSCAN method (orange ellipse in Figure 1c). The high efficiency of the NN method is probably related to the metric it uses, which is similar to the ETAS one with $\lambda_j(t_i, x_i) = (t_i - t_j)^{-1} r_{ij}^{-d_f} 10^{bm_j}$ and $c = 0$, $p = 1$, $d = 0$, $q = d_f$ and $a = b$. The windowing technique seems to overestimate the temporal and spatial windows, since it removes large amounts of seismicity (blue ellipse in Figure 1f), in accordance with previous results [11]. The same gap between the

14th and 15th year of the catalog is also evident in the background seismicity from the MAP-DBSCAN method, however, it is smaller and some sparse seismicity is left (blue ellipse in Figure 1c). On the other hand, Reasenberg's declustered catalog has more events than any other method (pink, magenta and green line, Figure 1a) and significant concentrations of events are visible in the space-time evolution of the background seismicity (orange and purple ellipses in Figure 1e).

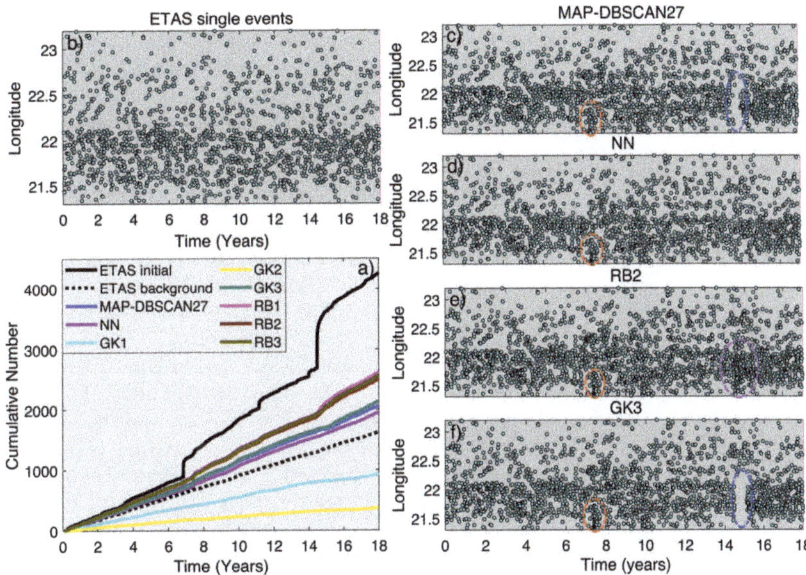

Figure 1. (**a**) Cumulative number of background events for each algorithm, the initial ETAS catalog with black color and the mother events of the ETAS synthetic catalog with the black dotted line. The space-time evolution (**b**) of the initial ETAS catalog and of the background seismicity for the four best algorithms, (**c**) MAP-DBSCAN27 ($J_2 = 0.647$), (**d**) NN ($J_2 = 0.727$), (**e**) RB2 ($J_2 = 0.630$), (**f**) GK3 ($J_2 = 0.676$). Colored ellipses stand for large gaps and significant concentration of events.

Best overlapping among the true, **X**, and the estimated partition, **Y**, does not mean necessarily the best detection for the declustered seismicity. For instance, the GK3 partition is characterized by a lower index, $J_1 = 0.585$, than the MAP-DBSCAN27 partition, $J_1 = 0.627$, however, its declustering catalog is more accurate (GK3-$J_2 = 0.676 > J_2 = 0.647$-MAP-DBSCAN27). Nevertheless, both indexes combined, the MAP-DBSCAN partition shows a higher efficiency than the rest of the algorithms, except for the NN. The Jaccard index values for the rest of the MAP-DBSCAN input parameters are quite stable with small fluctuations from the best parameter set (MAP-DBSCAN27), apart from the smallest distance cutoff, $\epsilon = 2.5$ km, which seems inadequate for capturing the spatial correlations among the events (Appendix B.4).

3. Earthquake Data

We considered three areas in the region of Greece (Figure 2a), which consist of distinctive seismotectonic units. The selection of the three study areas is based on criteria related to the homogeneity of the type of faulting, the comparatively intense continuous seismicity and the existence of seismic excitations during the study period. The first area, Corinth Gulf (CG) (Figure 2b), is undergoing high extensional deformation rates. The seismicity is mainly associated with eight major faults that bound the rift to the south and dip to the north [48]. The area of central Ionian Islands (CII) (Figure 2c) is characterized by the highest moment rate in the Mediterranean region. Its main seismotectonic feature is the Kefalonia

Transform Fault Zone (KTFZ) which extends for more than 100 km along the western coastlines of Lefkada and Kefalonia Islands, comprising two distinct main branches, the Lefkada and Kefalonia faults, respectively. Right lateral strike slip motion with a minor thrust component is the dominant faulting type [49,50]. The third area is located in the North Aegean Sea (NAS) (Figure 2d) and is dominated by dextral strike-slip faulting, along the North Aegean Trough (NAT) and its parallel branches [51], as a consequence of the westward propagation of the North Anatolian Fault into the Aegean [52]. The driving mechanism of the active deformation in the Aegean region is the subduction of the oceanic lithosphere of the Eastern Mediterranean under the continental Aegean microplate, forming the Hellenic Subduction Zone and the extensional back arc Aegean area due to the slab rollback [53].

For the investigation of the clustering properties in the three areas, we considered earthquake datasets from the regional catalog of the Geophysics Department of the Aristotle University of Thessaloniki [54], compiled with the recordings of the Hellenic Unified Seismological Network (HUSN). The earthquake catalogs of CG, CII and NAS, which we denote henceforth as D1, D2 and D3, include 25,595, 24,085 and 21,139 events, respectively, occurring between 2012 and 2019. For the determination of their completeness magnitude, we implemented the Goodness-of-Fit (GFT) method [55], assuming that earthquakes follow the Gutenberg–Richter (GR) law, $logN = a - bM$. In particular, the differences between the observed and the synthetic frequency-magnitude distributions are computed for increasing magnitude bins as threshold values. The completeness magnitude is defined as the first magnitude bin at which the difference falls under the 5% residual (Figure S1). Figure S1a–c show the residuals for the three datasets and Figure S1d–f present the GR law for the corresponding complete datasets. The b-value is calculated by means of the maximum likelihood method proposed by [56] and found equal to $b = 0.97, 0.88, 0.89$, for the D1, D2, D3 datasets, respectively (Table 3). The resulting magnitude threshold for the three datasets is equal to $M_c = 1.5, 2.2, 2.1$, with 13,043, 6981, 8328 events (Table 3), respectively, the epicenters of which are shown in the maps of each study area (Figure 2).

Table 3. The magnitude of completeness, M_c, for the datasets of the three areas CG, CII and NAS, along with the productivity, a, and the b-value of the GR law. N and N_c denote the initial number of events and the ones with $M \geq M_c$, respectively.

Region	Notation	N	M_c	N_c	a	b
CG	D1	25,595	1.5	13,043	5.57	0.97
CII	D2	24,085	2.2	6981	5.80	0.88
NAS	D3	21,139	2.1	8328	5.79	0.89

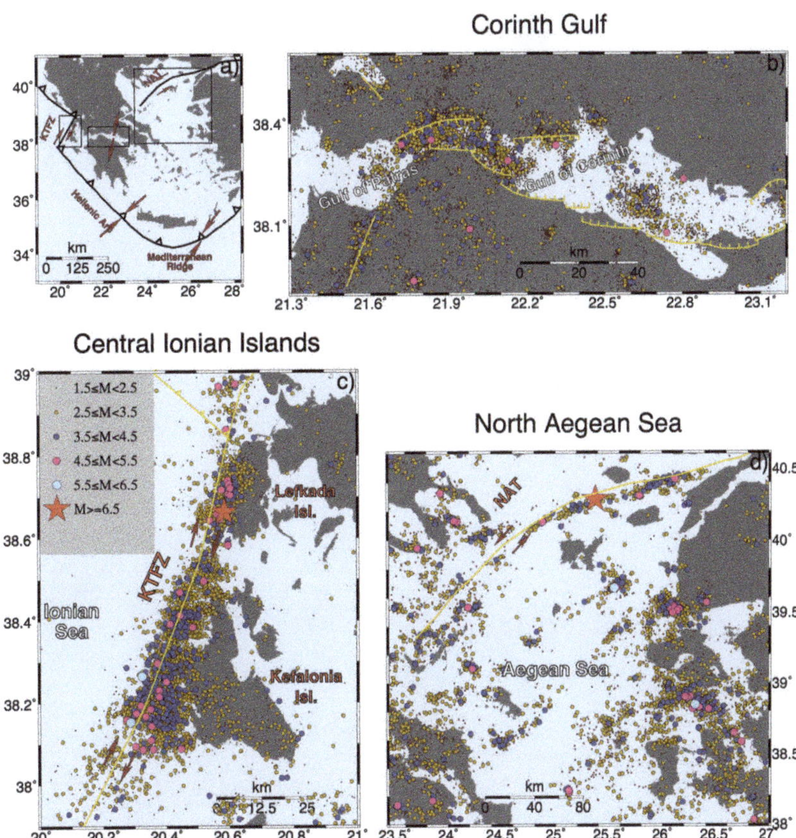

Figure 2. Maps of the study areas depicting seismicity along with major faults (yellow lines). (**a**) The main seismotectonic features of the area of Greece. The black lines illustrate the active boundaries and the arrows the relative plate motions. Rectangles enclose the three study areas. (**b**) The area of Corinth Gulf where the major faults are shown (yellow lines) along with seismicity during 2012–2019. (**c**) The area of Central Ionian Islands where the Kefalonia Transform Fault Zone and the collision front are shown (yellow lines) along with seismicity during 2012–2019. (**d**) The area of North Aegean Sea where the NAT is traced (yellow line) along with seismicity during 2012–2019. The legend is common for the three study areas.

4. Results

4.1. Triggered and Background Seismicity Separation

We fit MAP models with between two and seven states for each earthquake subcatalog, computationally a very demanding process as the number of states increases, especially for large datasets such as D3 with 13043 events. According to the BIC values, six, seven and again six states are sufficient to approximate the temporal distribution of earthquakes for the D1, D2 and D3 datasets, respectively. Next, we evaluate the transitions among the hidden states of the models. Firstly, the state probabilities $p_i(t) = p(J_t = i)$ for $i = 1, \ldots, K$ are estimated with the use of the forward and backward vectors given in Equation (A1), and then we assign as state of the hidden process J_t, the one with the highest probability, i.e., $\mathrm{argmax}_{0 \leq i \leq K} \, p_i(t)$, with $p_i(t) = p_i(t_k)$ for $t_k \leq t < t_{k+1}$. Each state i corresponds to an occurrence rate, λ_i, therefore, by evaluating the transitions among the states of the model, we detect change-points in the seismicity rate. Figure 3a–c illustrate the transitions among the states for the datasets D1, D2 and D3, respectively. The colored box at each temporal interval $t_k \leq t < t_{k+1}$ indicates the state with the maximum probability at the current time and the legend contains its corresponding occurrence rate.

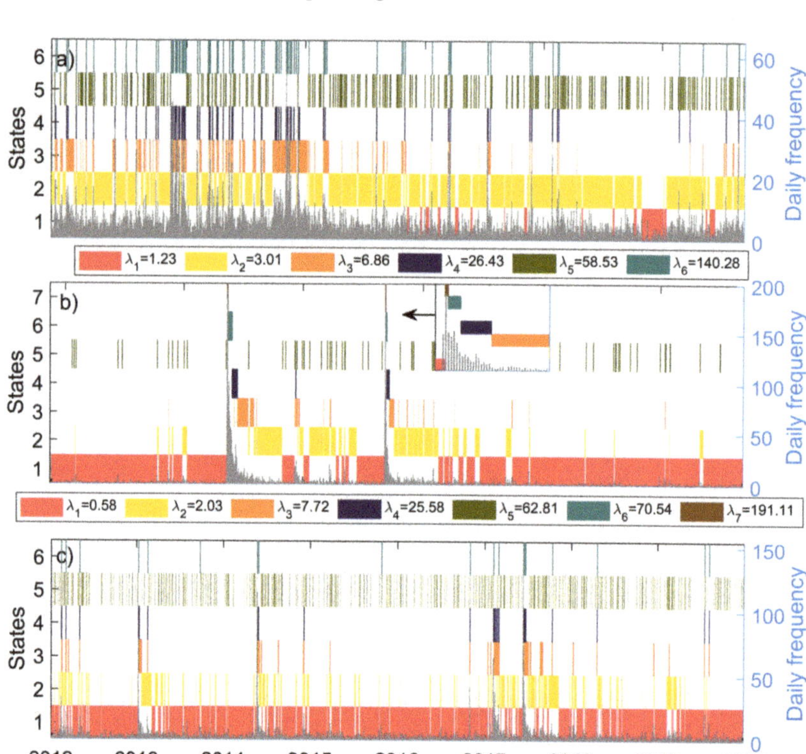

Figure 3. Most probable path of the hidden states of the model along with the daily frequency of events (gray vertical lines) with (**a**) $M \geq 1.5$ for D1, (**b**) $M \geq 2.2$ for D2 and (**c**) $M \geq 2.1$ for D3 datasets, respectively. Each color is assigned to a different state i with seismicity rate λ_i. Inset magnifies the transitions among the states, which are otherwise difficult to visualize due to the short sojourn times compared to the study period. The rate threshold, λ_{thr}, is set equal to $\lambda_2 = 3.01$, $\lambda_1 = 0.58$, $\lambda_1 = 1.53$ for the D1, D2 and D3 datasets, respectively.

The temporal patterns of dataset D1 indicate the dominance of state 2 (yellow color, Figure 3a) with occurrence rate $\lambda_2 = 3.01$ events/day for almost the entire period. Nevertheless, there is a slight decrease in the occurrence of earthquakes ($\lambda_1 = 1.23$ events/day) in the second part of the catalog, starting from 02/2016 with transitions to state 1 (red color, Figure 3a) until almost the end of the catalog in 12/2019. This is probably related to the lack of seismic sequences during the last part of the study period compared to the previous intense seismic activity especially during the period 2013–2014 in the western subarea of the CG [57]. The rate threshold is set equal to $\lambda_{thr} = \lambda_2$, which we consider as the background rate during the study period.

The seismicity of the CII area is dominated by the two major sequences during the study period, the 2014 Kefalonia doublet ($M_w6.1$ and $M_w6.0$) [58] and the $M_w6.5$ 2015 Lefkada earthquake sequence [59]. States 7 (brown), 6 (dark cyan), 4 (dark blue), 3 (orange) and 2 (yellow) in Figure 3b are clearly associated with the aftershock evolution of the two sequences—essentially, they approximate the Omori temporal distribution. Background seismicity is described by state 1 (red) with occurrence rate $\lambda_1 = 0.58$ events/day, which we set as rate threshold for the primary classification of the clusters.

Finally, dataset D3 also contains some major sequences, the 2013 $M_w 5.8$ [60], the 2014 $M_w 6.9$ Samothraki [61] and the 2017 $M_w 6.4$ Lesvos earthquake sequences [62], whose aftershock temporal distribution is approximated by states 6 (dark cyan), 4 (dark blue), 3 (orange) and 2 (yellow) of the model (Figure 3c). The rate threshold value is set equal to $\lambda_{thr} = \lambda_1$.

In general, we observe significant variations in the temporal evolution of the seismic excitations between the CG and the CII and NAS areas. Figure 3 illustrates that the daily frequency of events during seismic sequences in CII and NAS is decreasing in time, typical of mainshock–aftershock sequences, whereas in CG we observe large fluctuations in the daily frequency, common for earthquake swarms, as in 2014 when multiple seismic excitations occurred in the western subarea of CG.

Consecutive events above the rate threshold λ_{thr} are classified into groups which we call potential clusters, and then, we test four different sets of temporal constraints, (T, dt), to the three datasets. Potential clusters within a temporal interval T are merged into one and events that occurred in $\pm dt$ time from the potential cluster are also included. Next, the DBSCAN algorithm is implemented to the merged clusters in order to separate them based on their spatial density. The minimum number of neighbors for the determination of a cluster is set equal to 4 ($N_{pts} = 4$) for avoiding insignificant cases with fewer events. This is an appropriate choice for two-dimensional data according to Ester et al. [35]. For the determination of the distance threshold, ϵ, we computed the k-distances which is a procedure proposed by Ester et al. [35], which is commonly used to constrain the distance threshold [4]. In Appendix C we provide more details on the choice of the parameters and how they affect the spatio-temporal evolution of background seismicity.

4.2. Cluster Analysis

Table 4 gives the chosen parameter set of the clustering algorithm for each dataset based on the analysis in Appendix C and a summary on the statistics of the detected clusters. In the CG area, we identified the largest number of seismic clusters (255) due to the increased detectability of micro-seismicity (low completeness magnitude threshold), however, they are short in size ($\bar{n} = 18.28$) and duration ($\bar{\tau} = 12.50$) Conversely, the CII area is characterized by a small number of seismic clusters (45) but with large mean size ($\bar{n} = 118.43$) and duration ($\bar{\tau} = 54.60$). The clustered seismicity is prevalent (75%), whereas in CG and NAS, the background component is more dominant than clustered seismicity with 64% and 56%, respectively (Table 4). In CG, this is explained by the lack of large main shocks during the study period and the occurrence of few moderate events, the largest number with $M = 5.2$.

Concerning the CG area, the majority of the clusters are located on the western subarea where 22 out of 27 clusters with $N \geq 30$ occurred. The main activity is located offshore between Aigion and Trizonia Island, but also north of the Psathopyrgos fault (Figure 4a). The eastern subarea comprises smaller clusters that are mainly concentrated offshore Xylokastro and Perachora faults, as well as near Itea Gulf (Figure 5a). The seismicity of CII is dominated by the two major main shock–aftershock sequences, each sequence comprising 2829 and 1396 events, respectively. Essentially, 4225 out of the 5221 clustered events belong to these sequences (Table 4). Furthermore, 45 clusters are detected in total with the main activity concentrated along the KTFZ (Figure 6a). The NAS area comprises 187 clusters, including both main shock–aftershock sequences and earthquake swarms (Table 4). Figure 7a shows that the main clustered activity is concentrated along the NAT and the sub-parallel branches, as well as in the southeastern subarea.

Table 4. Cluster statistics and the parameter set of the clustering algorithm for the three datasets. N_{clust} corresponds to the number of clustered events and N_{bg} to the background seismicity frequency. $\bar{\tau}$ and $\bar{n}$ are the mean duration in days and size of the clusters, respectively.

Dataset	$(T, dt, \epsilon, N_{pts})$	N_{clust}	N_{bg}	# Clusters	$\bar{\tau}$	$\bar{n}$
D1	(0, 5, 2.5, 4)	4662 (36%)	8381 (64%)	255	12.50	18.28
D2	(5, 5, 2.5, 4)	5221 (75%)	1770 (25%)	45	54.60	118.43
D3	(5, 5, 5, 4)	3688 (44%)	4640 (56%)	187	15.08	19.72

4.2.1. Corinth Gulf Area

The western subarea of the Corinth Gulf is characterized by rich seismic activity, especially in 2013–2014, when 13 out of the 22 clusters with $N \geq 30$ occurred. One of the major detected sequences is the 2013 Aigion swarm (C6 in Figure 4b) which initiated on 21 May 2013 with a bulk of small events and several bursts associated to earthquakes with magnitudes ranging between 3.3–3.7 (Figure S3) [63,64]. Two distinct excitations followed (C8 and C10 in Figure 4b) in accordance with the ones observed by Michas et al. [65]. The first cluster began on 7 July with some activity prior to the $M = 3.7$ event on 15 July 2013, and lasted until 27 August, 2013 (Figure S3). The second half of 2014 is also a well-studied period with intense seismic activity. Five clusters with $N \geq 30$ are detected (C15, C16, C18, C19 and C20) in the western subarea, including the offshore M4.8 earthquake on 7 November 2014, associated with C19 (Figure 4b), and the M4.6 event on 21 September 2014, associated with the earthquake swarm located between Nafpaktos and Psathopyrgos [66] (C15 in Figure 4b). Persistent activity since 22 July 2014 is also observed offshore Aigion (C16), close to the earthquake swarm, C15, which began on 7 November 2014 (Figure S5). In 2012, fewer clusters are observed, mostly during the first semester, with three clusters comprising $N \geq 30$ events, C1, C2 and C3, and a plethora of smaller clusters (Figures 4b and S2). Between November 2013 and July 2014, the activity is sparse with three relatively large clusters, C11, C12 and C14 (Figures 4b and S4). Six more clusters with $N \geq 30$ are observed until the end of 2017 (C21, C22, C23, C24, C25, C27, Figure 4b).

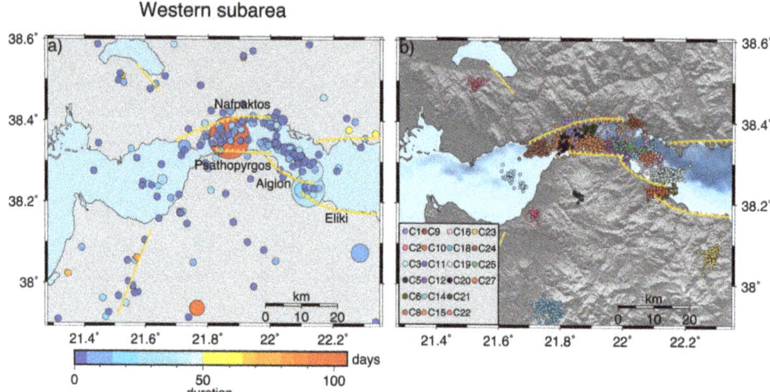

Figure 4. (**a**) Spatial distribution of the centroids of the identified clusters for the western subarea of Corinth Gulf along with major faults (yellow lines). The size of the circles is proportional to the earthquake number in each cluster, whereas the duration is represented by the color scale. (**b**) Spatial distribution of the clusters with $N \geq 30$ events. The index of each cluster is provided in the inset box.

The eastern subarea is characterized by more sparse activity. A major seismic sequence, Offshore Perachora (C4 in Figure 5b), is detected, including two sub-sequences, the first initiated on 22 September and the second on 30 September 2012 (Figure S6). Two relatively large clusters, *C13* and *C17*, are observed near Itea; the former lasted almost two weeks at the end of March, 2014, and the latter—almost three months between August and October 2014 (Figures 5 and S7).

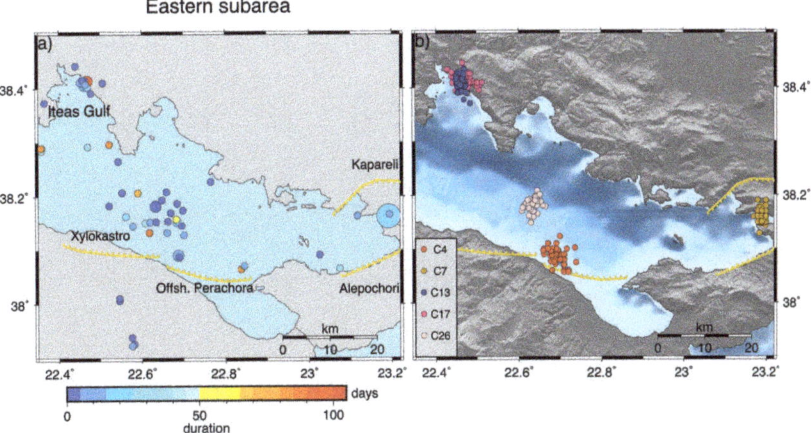

Figure 5. (**a**) Spatial distribution of the centroids of the identified clusters for the eastern subarea of the Corinth Gulf along with the major faults (yellow lines). The size of the circles is proportional to the earthquake number in each cluster whereas the duration is represented by the color scale. (**b**) Spatial distribution of the clusters with $N \geq 30$ events. The index of each cluster is provided in the inset box.

4.2.2. Central Ionian Islands Area

The two main shocks of sequence *I1* (Figure 6b) with $M = 6.1$ and $M = 6.0$ occupy the southern and the central part of the onshore area of Kefalonia Island. The 2014 Kefalonia earthquake sequence (*I1* Figure 6) started on 19 January with the first main shock occurring on 26 January ($M = 6.1$), and aftershock activity extending over 35 km [58], part of which hosted the second main shock ($M = 6.0$) that occurred on 3 February and the compound aftershock activity. A sub-cluster is also detected offshore to the southwest of Kefalonia Island (*I2* in Figure 6b) that is deployed concurrently with the main sequence (Figure S8). In addition, two distinct clusters, *I3* and *I4* (Figure 6b), are revealed, which occurred between November and December 2014 (Figure S9), across the edges of the double rupture. They might be triggered by the stress transfer of the main ruptures, indicating activation of adjacent fault segments. The seismic activity of cluster *I5* (Figure 6b) retains the most interest because it is essentially two seismic excitations evolving at the same time. The first initiated in the Myrtos Gulf and the second offshore the south part of Kefalonia Island. It comprises 164 earthquakes in about 100 days (Figure S9). The activity of the *I7* cluster (Figures 6b and S10) spreads along the western coastline of Lefkada and Kefalonia Islands, far beyond both sides of the 2015 Lefkada main rupture. To the south the aftershock activity is sparse, probably due to the large amount of stress released in the main rupture, revealing that the main slip is associated with a fault of about 17 km in length [59]. Apart from cluster *I4*, two additional clusters (*I6* and *I9* in Figure S9 and Figure S11, respectively) are detected in the area between Lefkada and Kefalonia, extending to about 15 km, which is considered as a transition zone encompassing step-over structures [58]. All of them relate to the E–W-oriented, parallel step-over faults, similar to the ones detected in the microseismicity cluster analysis between September 2016 and December 2019 in the study area [34].

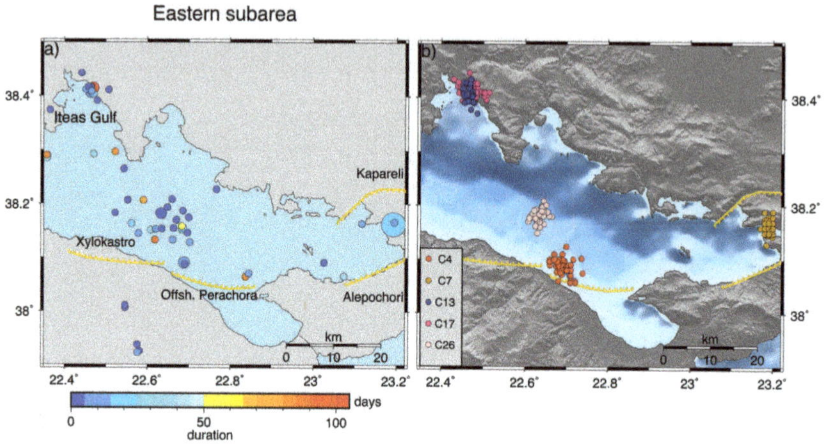

Figure 6. (**a**) Spatial distribution of the centroids of the identified clusters for the area of Central Ionian Islands along with the trace of the Kefalonia Transform Fault Zone (yellow lines). The size of the circles is proportional to the earthquake number in each cluster, whereas the duration is represented by the color scale. (**b**) Spatial distribution of the clusters with $N \geq 30$ events. The index of each cluster is given in the inset box.

4.2.3. North Aegean Sea Area

The first seismic excitation with $N \geq 30$ events ($N1$ in Figure 7b) is a sequence of interest since two moderate events ($M = 5.2$ and $M = 5.3$) occurred in 3 weeks, both producing their own aftershocks (Figure S12). The 2013, January 8 $M = 5.8$ North Aegean earthquake [60] along with its aftershock activity (cluster $N3$ in Figure 7b) is also detected. The aftershock activity is temporally divided into two clusters (Figure S13). The 24 May 2014 $M = 6.9$ Samothraki main shock was followed by aftershock activity confined to three major clusters ($N4$, $N5$, $N6$ in Figure 7b) and some secondary clusters with $N \geq 10$ events (Figure S14), which are in accordance with the ones observed by Saltogianni et al. [61]. The seismic activity that took place near the Aegean coast of NW Turkey during January–October 2017 [67] is divided into three clusters with $N \geq 30$ ($N10$, $N11$ and $N14$ in Figure 7b) and two minor clusters with 22 and 23 events, respectively (Figure S15). The strong main shock ($M = 6.4$) that occurred on the 12th of June 2017 offshore, south of the SE coast of Lesvos Island, along with its intense aftershock activity, is revealed and illustrated in Figure 7b ($N12$). Two major ($N \geq 30$) secondary outbursts of clustered activity occurred concurrently on the west ($N17$) and east ($N16$) side of the sequence (Figure S16). A thorough analysis revealing multiple spatial clusters of the sequences is conducted by Papadimitriou et al. [62].

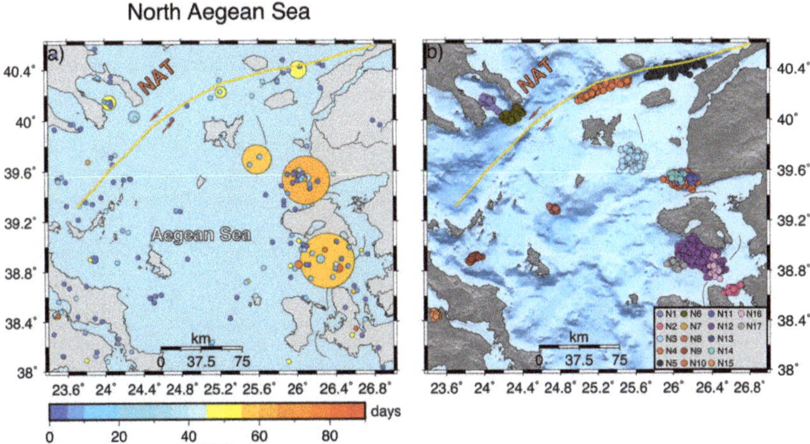

Figure 7. (**a**) Spatial distribution of the centroids of the identified clusters for the area of North Aegean Sea along with the trace of North Aegean Trough (yellow lines). The size of the circles is proportional to the earthquake number in each cluster, whereas the duration is represented by the color scale. (**b**) Spatial distribution of the clusters with $N \geq 30$ events. The index of each cluster is given in the inset box.

4.3. Regional Variability of Clustering Properties

In this section, we investigate regional variations in the clustering behavior of the detected seismic sequences, in particular, on their productivity rates and on their temporal evolution that can differ among areas with distinct seismotectonic characteristics. Therefore, we adopt the temporal ETAS model that expresses two empirical relationships that characterize the temporal and size distribution of earthquakes, the normalized Omori–Utsu law, given by $g(t) = c^{p-1}(p-1)(t-t_i+c)^{-p}$, and the productivity law that is expressed by $N = k(M_i) = Ke^{a(M_i-m_c)}$, where N is the number of triggered events by an earthquake of magnitude, M_i, K is a constant of proportionality, which depends on the number of triggered events per mainshock above the catalog cutoff, and a describes the impact of magnitude on the number of triggered events.

We compute the generic parameters for the 3 areas, CG, CII and NAS, by jointly inverting the ETAS parameter set $\theta = (p, c, a, K, \mu)$ from the identified sequences with $N \geq 30$ of each area. In particular, $LL_i = \sum_{j=1}^{n_i} \log \lambda(t_j) - \int_{t_0}^{t_{end}} \lambda(t)dt$ denotes the log-likelihood of the i-th sequence for each area, namely, the logarithmic probability of observing n_i events with occurrence times t_j, $j = 1, \ldots, n_i$, during the period of the sequence (t_0, t_{end}), and no other events between them. The intensity function, λ_t, of the model is given by Equation (1), neglecting the spatial component. We then stack all the sequences of each area, compute their corresponding logarithmic probabilities LL_i, and define as the common log-likelihood

$$LL = \sum_{i=1}^{N^*} LL_i, \tag{2}$$

where N^* is the number of sequences. The optimal inverted parameters are the ones that maximize Equation (2). The results of the ETAS parameter estimation for the three regions are shown in Table 5. There are 27 sequences in CG (Figures 4 and 5), 9 in CII (Figure 6) and 17 in NAS (Figure 7) from 2012 until 2019 with $N \geq 30$ events, however, we removed cluster C26 from the computations, since it is located at the boundaries of the study area (Figure 5) with part of the aftershock data being omitted. For the maximization of the common log-likelihood LL, we implement an iterative procedure where at each step we update the model parameters by a random factor so $\theta_k^{new} = \theta_k + u$, for $k = 1, \ldots, 5$, then, we compute the corresponding LL_i^{new}, $i = 1, \ldots, N^*$, log-likelihood values and store the new parameters under the condition $LL_{new} > LL$ moving to the next iteration. After some iterations, the logarithm converges and the algorithm stops. Essentially, this is a grid-based procedure, since we use a large number of iterations.

Table 5. Generic ETAS parameter values for the three study areas. N^* denotes the number of sequences with $N \geq 30$.

Area	p	c	a	K	μ	β	N^*
CG	1.23	0.0171	0.82	0.74	0.43	2.13	26
CII	1.31	0.11	1.29	0.44	0.15	2.21	9
NAS	1.26	0.0324	1.04	0.51	0.28	2.03	17

The parameter a for CG ($a = 0.82$) is the lowest among the three areas, indicating the dominance of swarm activity presumably due to fluid flow in accordance with many relevant studies [65,68]. Low a values characterize areas with high heat flow [39], even though the estimated value can be underestimated due to magnitude incompleteness after the occurrence of the main shock or due to the existence of time-dependent background seismicity [69]. Conversely, in CII, the estimated value ($a = 1.29$) is relatively larger compared to the former region ($a = 0.82$), indicating the dominance of typical main shock–aftershock sequences. In the NAS area, a moderate value is acquired ($a = 1.04$), probably due to the co-existence of swarm activity and aftershock sequences. Another indicator for the existence of swarm activity in CG is the large value of the background seismicity ($\mu = 0.43$) compared to NAS and CII. High values of the background rate can indicate the existence of aseismic loading transients [40]. LLenos et al. [70] observed increased values of the background component of the fitted ETAS model when it was applied to pre-swarm and swarm activity, respectively.

For the comparison of the productivity among the three areas, since they have different completeness magnitudes, we use the following relation,

$$N = k(M_i)P(M \geq m_c^*) = Ke^{a(M_i - m_c)}e^{-\beta(m_c^* - m_c)}, \tag{3}$$

which yields the number of earthquakes above magnitude m_c^*, generated by a main shock of magnitude M_i. Figure 8 shows the number of direct triggered events, $K(M_i)$, from an earthquake of magnitude M_i. We consider $m_c^* = 2.2$, which is the maximum completeness magnitude among the three datasets. The exponent of the exponential magnitude distribution is expressed by β and is defined as $\beta = \sum_{i=1}^{N} \beta_i / N^*$, where N^* is the number of clusters for each dataset and β_i their corresponding exponent values. Concerning the distribution of aftershocks in time, the normalized Omori law distribution is used, given by $g(t)$.

In CII, the seismic sequences seem to be more productive, as shown in Figure 8, with NAS and CG to exhibit smaller values. Combined with the higher background rate for the area of CG ($\mu = 0.43$), we may say that a significant part of Corinth Gulf's sequences cannot be contributed to the triggering effect of mainshocks but different underlying mechanisms seem to play an important role. Conversely, in CII area, mainshock–aftershock sequences seem to dominate, generating a rich number of aftershocks (very low background rate, $\mu = 0.15$, and high productivity of mother events).

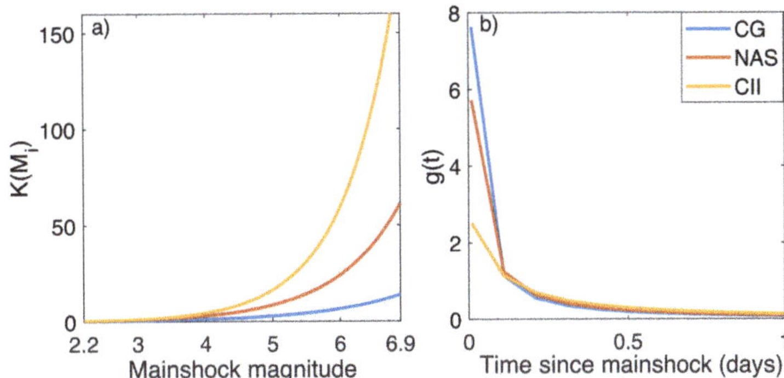

Figure 8. (**a**) The number of events triggered by an earthquake of magnitude M_i at CG (blue), NAS (orange) and CII (brown), respectively. (**b**) The temporal distribution of triggered aftershocks.

4.4. Sequence-Specific Clustering Properties

Next, we estimate the a-values for the individual sequences of each area by maximizing LL as a function of a, K and the background rate μ, while keeping the rest of the parameters fixed for clusters with $N < 80$. In this way, we increase the robustness of the inversion procedure since there are sequences with few events. A similar procedure was followed by Page et al. [3] and Llenos and Michael [36] who demonstrated that fitting multiple parameters for a single sequence can be unstable and Hardebeck et al. [37] who implemented this method for the estimation of California aftershock parameters. We intend to investigate potential differences in the productivity (a, K) and the background rate, μ, among sequences of each area and their relation to different underlying triggering mechanisms. Productivity parameters a and K are correlated, so we enabled both to run during the iterative procedure. We also examine the value of the background rate among sequences since it can be also an indicator of aseismic transients in a region. Both parameters, a and K, are not influenced by μ, as we verified it by implementing the inversion procedure, also keeping parameters a and K fixed.

4.4.1. Corinth Gulf

In Table 6, the inverted parameters for the 26 clusters of dataset D1 with $N \geq 30$ are given. We adopt the generic values of Omori law (p and c) for clusters with $N < 80$ to increase the stability of the inverted parameters. We observe relatively high background rates for most of the sequences and low a values, in particular, $a < 1$ for 10 out of the 26 clusters.

Concerning the 2013 Aigion earthquake swarm and its subsequent swarms (clusters C6, C9 and C10), we observe relatively low productivity values of the ETAS model ($a = 1.31$, 1.29, 1.13, Table 6) in accordance with studies suggesting pore-fluid pressure as the main triggering mechanism during the excitation [63]. Clusters C11 and C12 are part of the same swarm (Figure S4) that occurred offshore Psathopyrgos. Their relatively high background rates ($\mu = 1.34$, 1.92) show that a significant part of the clustered seismicity cannot be explained by the empirical laws of the triggering part of the ETAS model. Cluster 14 is part of a major swarm that began on 8 June 2014 (Figure S4). Michas et al. [65] did not find high diffusion rates that are related to fluid pore pressure. However, the large background rate found in our study ($\mu = 4.86$) and the low a value ($a = 0.92$) suggest the existence of a non-typical mainshock–aftershock sequence, with more complex triggering mechanisms being responsible, such as aseismic creep. Similarly, the largest cluster in the dataset, the C15, located offshore Nafpaktos, is characterized by relatively high background rate ($\mu = 1.32$) and low productivity ($a = 1.38$), more typical values for swarm activity. In contrast, clusters C18 and C19 that are more typical mainshock–aftershock sequences with a distinct in magnitude event in the initiation of the sequence (Figure S5), have low background rates ($\mu = 0.05$, 0.76) and relatively high productivity rates ($a = 1.77$, 1.80). The two clusters near Itea Gulf show contradictory results, in particular, the first one, C13, is characterized by a high background rate ($\mu = 3.41$), whereas the second, C17, which occurred four months later, exhibits a much smaller background value ($\mu = 0.35$) more typical for mainshock–aftershock sequences. However, biases can exist in the inversion of the parameters for clusters with a small number of events, so we should be cautious with the inference.

Table 6. Details on the 26 clusters with $N \geq 30$ events in CG area and the inverted ETAS parameters. The generic values of the Omori law, p and c, are adopted for clusters with $N < 80$.

ID	T_{in}	T_{end}	N	p	c	b	a	K	μ	M_{max}
C1	12/1/12	23/1/12	33	1.23	0.017	1.20	0.49	0.79	1.03	3.1
C2	13/1/12	27/1/12	33	1.23	0.017	0.83	1.69	0.23	1.25	3.1
C3	4/3/12	6/4/12	65	1.23	0.017	1.03	1.53	0.26	1.05	3.0
C4	22/9/12	3/10/12	69	1.23	0.017	0.99	0.36	0.94	1.44	5.0
C5	27/12/12	1/1/13	34	1.23	0.017	0.82	1.84	0.21	1.32	3.8
C6	22/5/13	28/6/13	310	1.45	0.012	0.96	0.20	0.90	0.47	3.7
C7	8/6/13	28/6/13	144	1.11	0.007	1.22	0.60	1.30	1.00	3.0
C8	7/7/13	27/7/13	128	1.04	0.001	0.77	0.34	2.48	0.59	3.7
C9	8/9/13	13/9/13	65	1.23	0.017	1.19	1.28	0.79	2.74	2.8
C10	29/10/13	6/11/13	68	1.23	0.017	1.27	0.10	0.91	2.87	3.1
C11	19/1/14	16/1/14	33	1.23	0.017	0.92	1.26	0.50	1.37	3.8
C12	29/1/14	10/2/14	70	1.23	0.017	0.81	1.39	0.29	1.92	3.9
C13	21/3/14	1/4/14	52	1.23	0.017	0.83	2.97	0.009	3.41	4.0
C14	8/6/14	11/6/14	74	1.23	0.017	0.81	0.92	0.64	4.86	4.3
C15	21/7/14	31/10/14	506	1.37	0.051	1.04	1.38	0.34	1.32	4.6
C16	22/7/14	1/11/14	95	1.26	0.014	1.15	0.72	0.45	0.44	2.8
C17	24/7/14	26/10/14	61	1.23	0.017	0.94	1.72	0.16	0.35	3.4
C18	23/7/14	31/10/14	121	1.25	0.131	0.95	1.77	0.24	0.05	4.7
C19	7/11/14	18/12/14	228	1.07	0.071	0.92	1.80	0.55	0.76	4.8
C20	7/11/14	14/12/14	36	1.23	0.017	1.05	1.27	0.41	0.42	3.1
C21	1/10/15	6/10/15	44	1.23	0.017	1.16	1.97	0.49	1.61	2.8
C22	27/7/16	5/8/16	32	1.23	0.017	0.75	3.50	0.09	0.45	2.7
C23	1/8/16	8/8/16	147	2.79	0.160	0.98	0.10	0.85	2.98	3.4
C24	9/1/17	23/1/17	104	2.79	0.702	0.82	1.70	0.15	1.05	4.5
C25	14/7/17	17/7/17	39	1.23	0.017	0.43	0.73	0.40	5.95	4.2
C27	30/10/17	2/11/17	31	1.23	0.017	0.50	1.68	0.10	6.19	3.5

4.4.2. Central Ionian Islands

In Table 7, the inverted parameters for the nine clusters identified in the area of CII with $N \geq 30$ events are given. We maintain fixed the Omori law parameters p and c (generic values) for clusters with $N < 80$. The estimated ETAS parameters of the sequence $I1$ are in accordance with the existence of a main shock–aftershock sequence described in Section 4.2. In particular, the background rate is relatively low ($\mu = 0.17$), indicating that the seismicity is adequately described by the triggering part of the ETAS intensity function. The seismic activity of clusters $I3$ and $I5$ (shown by green and blue color in Figure S9, respectively) are characterized by relatively high background rates ($\mu = 0.99$, 0.76, Table 7). The space-time evolution of the former indicates a rapid migration in the beginning of the sequence (Figure S9), whereas, for the latter, it is characterized by the smallest K value ($K = 0.04$) in the area although the a value is rather large. Taking into account the lack of distinct main shocks at the initiations of the sequences, they can be characterized as earthquake swarms, one of the few observed in an area which comprises mostly main shock–aftershock sequences. Concerning cluster $I9$, located in the transition zone between Lefkada and Kefalonia Islands, there is evidence for swarm activity due to the relatively high background seismicity rate ($\mu = 0.70$). Ultimately, the major main shock–aftershock sequences in the area, *I1*, *I7*, have the highest p values ($p = 1.42, 1.45$), meaning that they are characterized by high aftershock decay in time.

Table 7. Details on the 9 clusters with $N \geq 30$ events in CII area and the inverted ETAS parameters. The generic values of the Omori law, p and c, are adopted for clusters with $N < 80$.

ID	T_{in}	T_{end}	N	p	c	b	a	K	μ	M_{max}
I1	19/1/14	16/9/14	2829	1.42	0.24	0.79	1.31	0.40	0.17	6.1
I2	23/1/14	14/9/14	55	1.31	0.11	1.23	1.38	0.30	0.12	3.7
I3	5/11/14	11/12/14	134	1.36	0.06	0.99	1.44	0.29	0.99	5.1
I4	13/11/14	12/12/14	66	1.31	0.11	0.93	1.43	0.38	0.37	4.9
I5	5/1/15	27/4/15	164	1.05	0.01	0.93	2.82	0.10	0.76	4.4
I6	18/1/15	24/4/15	71	1.31	0.11	1.08	1.91	0.36	0.15	3.8
I7	13/11/15	26/6/16	1396	1.45	0.30	0.86	1.51	0.29	0.45	6.5
I8	20/11/15	25/6/16	65	1.31	0.11	0.84	0.94	0.53	0.07	4.3
I9	4/4/17	4/5/17	67	1.31	0.11	0.95	2.26	0.18	0.70	3.9

4.4.3. North Aegean Sea

In NAS area cluster $N1$, which consists of two moderate events ($M = 5.2$ and $M = 5.3$) within a time period of 3 weeks, exhibits the lowest a value ($a = 1.10$) among the main detected clusters, which could be an indicator of fluid diffusion in the area (Table 8). Another case worth mentioning is the 24 May 2014, $M = 6.9$, Samothraki seismic sequence which is divided into three major clusters ($N4$, $N5$, $N6$, in Figure 7). The estimated background rates of the three major clusters are relatively small ($\mu = 0.16, 0.60, 0.29$), whereas the opposite holds for the scaling parameter, a, for the first two clusters ($a = 1.82, 1.76$). Concerning the seismic excitation that consists of clusters $N10$, $N11$ and $N14$, the relatively low productivity rates of the ETAS model ($a = 1.31, 1.29, 1.13$) and, conversely, the relatively high background rates for the first two, $N10$ and $N11$, clusters ($\mu = 1.00, 0.91$) may indicate fluid intrusion. This observation is in accordance with the study of Mesimeri et al. [67] who derived high background rates after the estimation of the ETAS model to the empirically divided 5 sub-clusters of the primary seismic activity (January–March 2017). A fast-diminishing aftershock activity is observed for the main shock ($M = 6.4$) that is located SE of Lesvos Island ($N12$), which is translated into a high Omori exponent, $p = 1.48$. Additionally, low background rates characterize the three main clusters, $N12$, $N16$ and $N17$, indicating that they are probably related to tectonic and coseismic stress transfer from previous seismicity [62]. Worth mentioning are the remarkable high background rates for clusters $N8$ ($\mu = 1.76$) and $N9$ ($\mu = 2.78$), which could be an indicator for seismic activity driven by transient forces, however, the number of events is rather small and could have led to significant biases in the inversion of the parameters.

5. Discussion

The consistency and efficiency of the MAP-DBSCAN method is examined on a simulated earthquake catalog of 18 years that produces the main features of seismicity in the region of Greece. In particular, we showed that our method is able to identify the connections among the events generated by a spatiotemporal ETAS model, as well as the mother events that initiated each cluster. The knowledge of the links among the events enabled the comparison of the method with some well known clustering algorithms, like the Gardner and Knopoff, the Reasenberg and the Nearest-Neighbor, by the use of the Jaccard index. This is a tool for measuring the overlap between the original partition of events into clusters and background seismicity, and the estimated one after the implementation of each clustering method. The results show that MAP-DBSCAN method is very competitive and in most cases outperforms the tested algorithms. The NN achieves the best reconstruction of the clusters (Table 2), which is probably related to the similarity of its metric with the ETAS metric that is used for the generation of the seismicity. The window-based method overestimates the clustered seismicity in accordance with work by Peresan and Gentili [11], whereas the Reasenberg link-based method seems to overestimate the background events (Figure 1).

The advantage of using the MAP model lies in its efficiency in capturing the changes in seismicity rate, independently of the mechanisms responsible for each seismic sequence. Furthermore, in case of non-stationary background seismicity, the MAP model can approximate the different phases by embedding multiple states into the Markov process J_t, i.e., distinct occurrence rates, and adopting a multiple rate threshold alternating according to the phase of the process each time. In this way, although it is more complicated, we can model both the non-stationary background seismicity and the triggered events without declustering the earthquake catalog [33]. The DBSCAN algorithm does not assume any specific spatial distribution of earthquakes and settles them into groups based solely on their spatial density.

We applied our method to three seismic zones in Greece during 2012–2019, identifying the major seismic sequences and a plethora of smaller ones. The rich seismic activity during 2013–2014 in the western subarea of the Corinth Gulf is detected in detail, a nontrivial issue, especially for the area between Nafpaktos-Psathopyrgos and offshore Aigion, where multiple excitations occurred in close proximity and within short periods (Figures 4, S4 and S5). Seismicity in the eastern subarea of the Corinth Gulf is found to be more sparse with few major clusters located near Itea Gulf (Figures 5 and S7) and offshore Perachora and Xylokastro (Figure 4 and S6). On the contrary, seismicity in the Central Ionian Islands is dominated by the two major main shock–aftershock sequences associated with the 2014 Kefalonia and the 2015 Lefkada seismic sequences (Figure 6). Together they comprise the 81% of the clustered seismicity in this area. Many large clusters are identified in the North Aegean Sea area that includes both main shock–aftershock sequences and earthquake swarms.

We investigated the properties of clustering seismicity among the three study areas with the use of the ETAS model. The results indicate that there are differences in aftershock productivity rates between Corinth Gulf, Central Ionian Islands and North Aegean Sea, showing that productivity can vary regionally. As showed by Page et al. [3] and LLenos and Michael [36] adopting the regional variations of productivity can produce a significant gain on aftershock forecasts. In the Central Ionian Islands, main shock–aftershock sequences seem to be more productive with the North Aegean Sea and the Corinth Gulf to follow (Figure 8). The sequences in the Corinth Gulf in particular are characterized by the highest background rate among the three areas (Table 5), meaning that a significant portion of clustered seismicity is not caused by the triggering of a main shock coseismic slip, but by the contribution of different triggering mechanisms. Many studies have focused on this area, suggesting pore-pressure changes due to fluid migration and aseismic creep as possible triggering mechanisms for the clustered seismicity [57,71]. In the North Aegean Sea, the swarm activity coexists with aftershock sequences, implying that for forecasting purposes, a finer regionalization might be more appropriate.

We also investigated potential differences in the productivity and the background rates among sequences of each region and their relation to different underlying triggering mechanisms. Results show that the high background seismicity (μ) and low productivity (a) values of the ETAS model are related to earthquake swarm activity triggered by fluid pore-pressure changes, such as the 2013 Aigion swarm (clusters $C6$, $C9$ and $C10$, Table 6, Figures 4 and S3) in Corinth Gulf [63] and the 2017 Tuzla earthquake swarm (clusters $N10$, $N11$ and $N14$, Table 8, Figures 7 and S15) in North Aegean Sea [67]. This is in accordance with studies suggesting the dependence of low productivity values to the existence of fluids [39,69]. In general, 18 out of 26 clusters in Corinth Gulf have background rates $\mu > 1$ and low productivity values (11 out of 26 with $a < 1$), whereas in the Central Ionian Islands, where main shock–aftershock sequences dominate, we observe very low background rates of the ETAS model (all with $\mu < 1$) and relatively high productivity values. In the North Aegean Sea area, we cannot observe a clear pattern, however, the majority of the detected clusters are characterized by low background rates and relatively high productivity, suggesting the dominance of typical main shock–aftershock sequences.

Table 8. Details on the 17 clusters with $N \geq 30$ events in NAS area and the inverted ETAS parameters. The generic values of the Omori law, p and c, are adopted for clusters with $N < 80$.

ID	T_{in}	T_{end}	N	p	c	b	a	K	μ	M_{max}
N1	14/2/12	4/4/12	136	1.41	0.03	0.96	1.10	0.40	0.54	5.3
N2	27/4/12	3/5/12	30	1.26	0.03	0.53	1.64	0.16	1.07	4.8
N3	8/1/13	6/3/13	285	1.07	0.06	0.88	2.39	0.06	0.55	5.8
N4	24/5/14	9/7/14	94	1.41	0.78	0.74	1.82	0.01	0.16	6.9
N5	24/5/14	11/7/14	153	1.60	0.16	0.69	1.76	0.16	0.60	4.5
N6	24/5/14	22/6/14	83	1.49	0.04	0.64	1.25	0.30	0.29	4.4
N7	6/12/14	29/12/14	41	1.26	0.03	0.67	1.60	0.15	0.31	4.9
N8	26/3/15	2/4/15	30	1.26	0.03	0.97	1.45	0.36	1.76	4.1
N9	29/10/16	31/10/16	49	1.26	0.03	0.89	2.44	0.28	2.88	3.4
N10	26/1/17	28/3/17	568	1.29	0.04	0.73	1.31	0.36	1.00	5.1
N11	7/4/17	12/5/17	38	1.26	0.03	1.05	1.29	0.11	0.91	3.4
N12	12/6/17	8/8/17	614	1.48	0.12	0.79	1.46	0.25	0.86	6.4
N13	13/6/17	29/7/17	48	1.26	0.03	1.03	2.42	0.17	0.35	3.7
N14	15/8/17	23/10/17	38	1.26	0.03	1.06	1.13	0.39	0.26	3.5
N15	16/8/17	11/11/17	34	1.26	0.03	1.08	1.46	0.36	0.15	3.5
N16	17/8/17	8/11/17	39	1.26	0.03	1.24	2.39	0.14	0.31	3.2
N17	24/8/17	11/11/17	35	1.26	0.03	1.01	2.23	0.13	0.27	3.6

6. Conclusions

In this study, we present the efficiency of our clustering method, MAP-DBSCAN, on a simulated earthquake catalog where the structure of the clusters is known a priori and its competitiveness against well-known clustering algorithms, as in most cases, shows better results. The main seismic clusters in the Corinth Gulf, Central Ionian Islands and North Aegean Sea during 2012–2019 are detected by our method and their clustering properties are investigated. The results show the existence of regional variability in aftershock productivity and background rates. In particular, the Corinth Gulf is characterized by low productivity values and high background rates related to the dominance of earthquake swarms, whereas seismicity in the Central Ionian Islands is comprised by main shock–aftershock sequences with high productivity. Sequence-specific parameters verify the dependence between low productivity values and high background rates with pore-pressure due to fluids migration. We believe that future studies on Operational Earthquake Forecasting should incorporate localized parameters into the models to improve the forecasting accuracy.

Supplementary Materials: The following are available online at https://www.mdpi.com/article/10.3390/app12041908/s1, Figure S1a–c Residuals (purple triangles) as a function of minimum cutoff magnitude, M_c, for the D1, D2 and D3 datasets, respectively. Blue and cyan dotted horizontal lines indicate the 10% and 5% residual thresholds, respectively. M_c (red triangle) is found as the first magnitude cutoff at which the confidence 95% is reached. (d–f) Incremental (red triangles) and logarithmic cumulative frequency (blue triangles) as a function of magnitude. The black line is the GR law fit according to the GFT method with $M_c = 1.5, 2.2, 2.1$ for datasets D1, D2 and D3, respectively. Figure S2 (a) Epicentral map of the main seismic clusters during the first semester of 2012. Three major clusters, C1, C2 and C3, and eight smaller clusters with $N \geq 10$ events occurred. (b) Space-time evolution of seismicity. Colors correspond to different clusters and the size of circles is proportional to the earthquakes' magnitude, Figure S3 (a) Epicentral map of the 2013 Aigion swarm and subsequent sequences in the area with $N \geq 10$ events. (b) Space-time evolution of seismicity. Colours correspond to different clusters and the size of circles is proportional to the earthquakes' magnitude, Figure S4 (a)

Epicentral map of the seismic activity between November, 2013 and June, 2014. Twelve clusters with $N \geq 10$ occurred, including the $C11$, $C12$ and $C14$ clusters. (b) Space-time evolution of seismicity. Colors correspond to different clusters and the size of circles is proportional to the earthquakes' magnitude, Figure S5 (a) Epicentral map of the intense seismic activity during the second half of 2014. Five major clusters occurred, the $C15$, $C16$, $C18$, $C19$ and $C20$, and four smaller clusters with $N \geq 10$ events. (b) Space-time evolution of seismicity. Colors correspond to different clusters and the size of circles is proportional to the earthquakes' magnitude, Figure S6 (a) Epicentral map of the seismic sequence Offsh. Perichora. One major cluster, $C4$, including two sub sequences, the first initiated on 22 September and the second on 30 September, 2012. b) Space-time evolution of seismicity. Colours correspond to different clusters and the size of circles is proportional to the earthquakes' magnitude, Figure S7 (a) Epicentral map of the seismic activity near Itea Gulf during 2014. Two major clusters are occurred, the $C13$, $C17$ and four smaller ones with $N \geq 10$ events. (b) Space-time evolution of seismicity. Colors correspond to different clusters and the size of circles is proportional to the earthquakes' magnitude, Figure S8 (a) Epicentral map of the 2014 Kefalonia earthquake sequence, $I1$, and a sub-cluster, $I2$, that occurred offshore the southern part of Kefalonia Island. (b) Space-time evolution of seismicity. Colors correspond to different clusters and the size of circles is proportional to the earthquakes' magnitude, Figure S9 (a) Epicentral map of four main clusters, $I3$, $I4$, $I5$ and $I6$ with $N \geq 30$ between November, 2014 and April, 2015. (b) Space-time evolution of seismicity. Colors correspond to different clusters and the size of circles is proportional to the earthquakes' magnitude, Figure S10 (a) Epicentral map of the 2017 Lefkada sequence, $I7$, along with two sub-clusters in the southwestern part of Kefalonia Island. (b) Space-time evolution of seismicity. Colors correspond to different clusters and the size of circles is proportional to the earthquakes' magnitude, Figure S11 (a) Epicentral map of cluster $I9$ located in the area between Lefkada and Kefalonia. Right: Space-time evolution of seismicity. Colors correspond to different clusters and the size of circles is proportional to the earthquakes' magnitude, Figure S12 (a) Epicentral map of cluster $N1$ comprised by two sub-sequences. (b) Space-time evolution of seismicity. Colors correspond to different clusters and the size of circles is proportional to the earthquakes' magnitude, Figure S13 (a) Epicentral map of the 2013 North Aegean sequence, denoted $N3$. (b) Space-time evolution of seismicity. Colors correspond to different clusters and the size of circles is proportional to the earthquakes' magnitude, Figure S14 (a) Epicentral map of the 2014, Samothraki sequence confined into three major clusters, $N4$, $N5$ and $N6$. (b) Space-time evolution of seismicity. Colors correspond to different clusters and the size of circles is proportional to the earthquakes' magnitude, Figure S15 (a) Epicentral map of the seismic activity near the Aegean coast of NW Turkey during January–October 2017 confined into three clusters, $N10$, $N11$ and $N12$. (b) Space-time evolution of seismicity. Colors correspond to different clusters and the size of circles is proportional to the earthquakes' magnitude, Figure S16 (a) Epicentral map of the 2017 sequence ($N12$) that occurred offshore, south of the SE coast of Lesvos Island along with its intense aftershock activity. Two major secondary bursts of activity occurred concurrently on the west ($N17$) and east ($N16$) side of the sequence. (b) Space-time evolution of seismicity. Colors correspond to different clusters and the size of circles is proportional to the earthquakes' magnitude.

Author Contributions: Conceptualization, P.B. and E.P.; methodology, P.B.; software, P.B.; validation, P.B.; formal analysis, P.B. and G.T.; data curation, P.B. and E.P.; writing—original draft preparation, P.B.; writing—review and editing, E.P. and G.T.; visualization, P.B.; supervision, E.P. and G.T.; funding acquisition, P.B. All authors have read and agreed to the published version of the manuscript.

Funding: This research is co-financed by Greece and the European Union (European Social Fund-ESF) through the Operational Programme Human Resources Development, Education and Lifelong Learning in the context of the project "Strengthening Human Resources Research Potential via Doctorate Research" (MIS-5000432), implemented by the State Scholarships Foundation (IKY).

Institutional Review Board Statement: Not applicable.

Informed Consent Statement: Not applicable.

Data Availability Statement: The data presented in this study are openly available at https://doi.org/10.7914/SN/HT (accessed on 15 January 2021).

Acknowledgments: The editorial assistance and the constructive comments from two anonymous reviewers are greatly appreciated. We are also grateful to I. Zaliapin for providing the code for nearest-neighbor analysis. The software Generic Mapping Tools was used to plot the map of the study area [72]. Geophysics Department Contribution 959.

Conflicts of Interest: The authors declare no conflict of interest.

Appendix A

The estimation of the MAP parameter set $\theta = \{\lambda_i, q_{ij}\}$ is based on the maximization of its likelihood function, $L(\theta|T) = \pi_{arr}^T e^{\mathbf{D}_0 \tau_1} \mathbf{D}_1 \ldots e^{\mathbf{D}_0 \tau_N} \mathbf{D}_1 \mathbf{1}_K$, with interevent times, $\tau_i = t_{i+1} - t_i, i = 1, \ldots, N$, comprising the trace $T = \{\tau_1, \ldots, \tau_N\}$ of the data, K hidden states and $N + 1$ number of events. The EM algorithm [73] is used for the optimization of the likelihood function, which is a common procedure for applications with hidden data, i.e., the change points of seismicity rate. At each iteration of the algorithm, the log-likelihood (LL) function is computed through the forward and backward vectors, which describe the evolution of the process recursively. The i-th element of the forward vector $\mathbf{f}[k] = \{f_i(k), i = 1, \ldots, K\}$ gives the probability to be in state i by taking into account the history of occurrences up to time t_{k+1}. Their values are obtained recursively through $\mathbf{f}[k]_j = \sum_{i=1}^K \mathbf{f}[k-1]_i e^{-\lambda_i \tau_k} q_{ij}(1)$, with $\mathbf{f}[0] = \pi_{arr}$. Similarly, the backward vectors $\mathbf{b}[k] = \{\mathbf{b}_i(k), i = 1, \ldots, K\}$ are defined giving the likelihood function $L(\theta|T) = \mathbf{f}[k]\mathbf{b}[k+1]$. Additionally, the forward and backward equations are used for the evaluation of the transitions among the states of the Markov process J_t. This is crucial for the implementation of our method, since it allows the detection of changes in the seismicity rate. The state probabilities of the hidden process J_t at a given time t_k are obtained by

$$p_{t_k}(i) = P(J_{t_k} = i) = \frac{p(\tau_1, \ldots, \tau_N, J_{t_k} = i)}{p(\tau_1, \ldots, \tau_N)} = \frac{\mathbf{f}[k-1]_i \mathbf{b}[k]_i}{L(\theta|T)}, \tag{A1}$$

For the determination of the hidden states number that is appropriate for capturing the seismicity rate changes, the Bayesian Information Criterion (BIC) [74] is used, which is a metric based on the maximum log-likelihood of each model. It is expressed through $BIC = -2 \times LL + log(N^*) \times k$, where k is the number of estimated parameters and N^* corresponds to the number of observations.

Appendix B

Appendix B.1. Gardner and Knopoff Window-Based Method

The procedure introduced by Gardner and Knopoff [10] for the detection of aftershocks is based on specific magnitude dependent space-time windows. It is known as the window-based method, and it is one of the simplest forms of aftershock identification. For each earthquake with magnitude M, the subsequent events are assigned as aftershocks if they occur within a temporal window $t(M)$ and a spatial interval $d(M)$, respectively. Foreshocks are treated as aftershocks when a larger earthquake occurs later in the sequence. The event is considered as an aftershock and the algorithm is repeated based on the largest magnitude.

We give in Equation (A2) the functional form of the spatial and temporal windows suggested in Gardner and Knopoff [10], which are denoted as GK1. Additionally, in Equations (A3) and (A4) we present alternative window parameter settings that can be found in van Stiphout et al. [75]. We denote them as GK2 and GK3, respectively.

$$d = 10^{0.1238*M+0.983} \text{ (km) and } t = \left\{ \begin{array}{ll} 10^{0.032*M+2.7389} & M \geq 6.5 \\ 10^{0.5409*M-0.547} & M < 6.5 \end{array} \right\} \text{ days} \tag{A2}$$

$$d = e^{1.77+\sqrt{0.037+1.02*M}} \text{ (km) and } t = \left\{ \begin{array}{ll} 10^{2.8+0.024*M} & M \geq 6.5 \\ e^{-3.95+\sqrt{0.62+17.32*M}} & M < 6.5 \end{array} \right\} \text{ days} \tag{A3}$$

$$d = e^{-1.024+0.804*M} \text{ (km) and } t = e^{-2.87+1.235*M} \text{ days} \tag{A4}$$

Appendix B.2. Reasenberg Linked-Based Method

In this method, an interaction zone among earthquakes is assumed that is modeled based on estimates of the stress redistribution for the spatial extent and on a probabilistic model, the Omori law, for the temporal extent, respectively. Any earthquake that occurs within the interaction zone of a prior earthquake is considered an aftershock and is included in the cluster. For the Reasenberg algorithm, we used the ZMAP tool [76] and we adopted 3 different sets of parameters given in Table A1. The parameters τ_{min} and τ_{max} correspond to the minimum and maximum elapsed time since the last event, in order to observe the next correlated earthquake at a certain probability, p_1. Additionally, x_{meff} denotes the minimum magnitude threshold for the earthquake catalog, whose value in the clusters is raised by a factor x_k of the largest earthquake within. Finally, the parameter r_{fact} corresponds to the radii we adopt to consider linking a new event with the cluster.

Table A1. Input parameters for the Reasenberg clustering algorithm. The first row corresponds to the standard parameter set [12].

PS	τ_{min}	τ_{max}	p_1	x_k	x_{meff}	r_{fact}
RB1	1	10	0.95	0.5	2.5	10
RB2	1	10	0.95	0.5	2.5	20
RB3	0.5	20	0.95	0.5	2.5	20

Appendix B.3. Nearest-Neighbor Method

The approach is based on the space-time-magnitude distance metric among two earthquakes given by Baiesi and Paczuski [22]:

$$\eta_{ij} = (t_j - t_i) r_{ij}^{d_f} 10^{-bm_i}, \tag{A5}$$

where r_{ij} is the epicentral distance between events i and j, d_f is the spatial fractal dimension and b is the component of the Gutenberg–Richter distribution. Each event j is connected to its nearest neighbor $i = argmin_{i:t_j>t_i} \eta_{ij}$ if their distance, η_j, is lower than a predefined threshold η_0. The earthquake catalog is then partitioned on distinct clusters, each containing at least one event. For the selection of the threshold value, η_0, the logarithm of the nearest neighbor distance $\eta^* = \{\eta_j\}_{j=1,...,N}$ is considered, where N the number of events. It follows an 1D Gaussian distribution with two components, which is essentially a mixture model of two Gaussian densities with parameters $N(\mu_1, \sigma_1)$, $N(\mu_2, \sigma_2)$ and a_1, a_2 weights, respectively. Then, the intersection of the two functional forms gives the threshold value.

There are only two free parameters, the fractal dimension d_f and b value, which are considered equal to $d_f = 1.51$ and $b = 1.0$, respectively. The logarithm of the separation distance is equal to $\log \eta_0 = -5.04$, based on the intersection of the two modes in the 1D density distribution of distances (Figure A1).

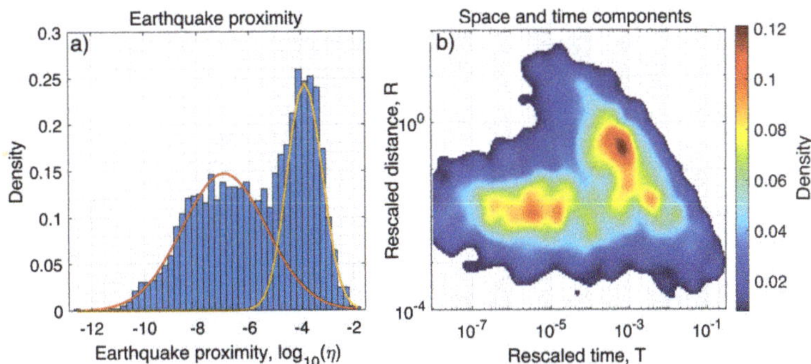

Figure A1. Distribution of the NN distances among all pairs of earthquakes of the ETAS synthetic catalog. (**a**) 1D density distribution of $\log \eta$, with estimated Gaussian densities for clustered (yellow) and background (orange) components. (**b**) 2D joint distribution of rescaled space and time distances.

Appendix B.4. MAP-DBSCAN Method

A MAP with 7 states is chosen based on BIC, the rate threshold is set to $\lambda_{thr} = \lambda_1$, and different temporal windows are tested for merging the potential clusters. Finally, the DBSCAN algorithm is implemented for 5 different distance thresholds (ϵ). The minimum number of events is set to $N_{pts} = 2$ for a better comparison with the other methods where clusters with at least 2 events can be defined. In Table A2 we present details on the parameter tuning.

Table A2. The 30 different parameter sets used for the detection of the clusters.

ϵ	N_{pts}	PS	T	dt	PS	T	dt
		1–5	0	0	16–20	0	7
		6–10	7	0	21–25	7	7
[2.5 5 7.5 10 12.5]	2	11–15	14	0	26–30	14	7

The method seems rather insensitive to the parameter selection. In particular, Figure A2 presents the Jaccard index values that describe the efficiency of the method to correctly reconstruct the initial clusters (J_1) as well as to identify the single events (J_2). We observe that the Jaccard index values are quite stable with small fluctuations, apart from the smallest upper-distance cutoff, $\epsilon = 2.5$ km, which seems inadequate to capture the spatial correlations among the events. Furthermore, the contribution of the temporal constraints to the clustering procedure seems negligible, with the exception of the two peaks for PS12 and PS27. This is an indicator that the MAP model has already achieved a sufficient separation between background and triggered seismicity based on the embedded multiple rates of the model.

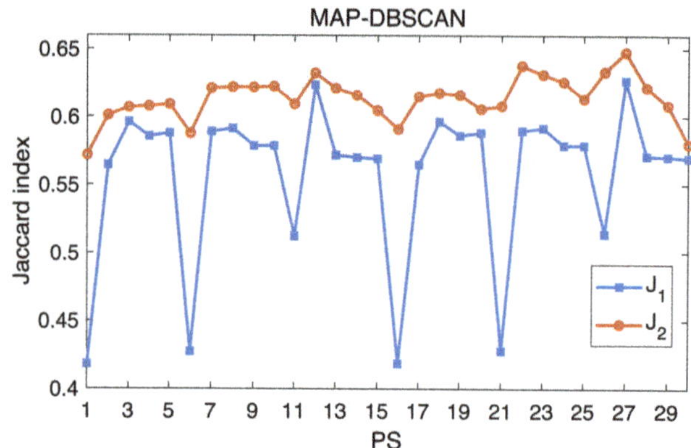

Figure A2. The Jaccard index values, J_1 with blue and J_2 with orange color, respectively, for all the input parameters of the MAP-DBSCAN method.

Appendix C

We implemented the clustering procedure MAP-DBSCAN for 16 different combinations of parameters which are shown in Table A3. For the determination of the distance threshold ϵ, we computed the k-distances between events assigned to the same potential cluster, since the DBSCAN algorithm is implemented in events that have been already grouped into clusters based on their temporal proximity. In particular, for each event included in the potential cluster, its k-nearest neighbor is computed and plotted in ascending order. If we choose an arbitrary event, i, set the distance threshold ϵ to k-dist(i) and the parameter N_{pts} to k, all events with an equal or smaller k-dist value will become core points, in other words, they will be assigned into a cluster. Ester et al. [35] proposed as best ϵ value the one that corresponds to a change in the slope of the curve, as corner points indicate a change in the degree of correlation among events. For $k = 4$, which corresponds to the minimum number of neighbors (N_{pts}), gradient changes in the slope range between 2.5 and 10 km in the datasets of both CG (Figure A3a) and NAS (Figure A3c) areas, whereas for the CII area (Figure A3b), changes in the slope of the curves initiate slightly sooner (below 2.5). The minimum one is chosen as equal to $\epsilon = 2.5$ in order to also ensure that the location errors of the catalog are considerably fewer.

Table A3. The 16 tested parameter of MAP-DBSCAN method for the three datasets D1, D2 and D3.

ϵ	N_{pts}	PS	T	dt	PS	T	dt
[2.5 5 7.5 10]	4	1	0	0	3	0	5
		2	5	0	4	5	5

For the 16 different realizations of the clustering algorithm, MAP-DBSCAN, we investigated the spatio-temporal properties of the background seismicity. Figure A4 presents the cumulative number of events that have not been assigned to a cluster (declustered seismicity) for each set of parameters along with the initial datasets. Peaks and pronounced concavities in the cumulative curves are indicators of triggered seismicity wrongly assigned as background and vice versa. In datasets D1 and D2 we observe such concaves for thresholds $\epsilon \geq 5$ km and a rather stable curve for $\epsilon = 2.5$ km (Figure A4a–h), suggesting that events are correctly separated as background and triggered ones. Therefore, the distance threshold is set to $\epsilon = 2.5$ km, for both datasets. In dataset D3, Figure A4i–l show that the curves with $\epsilon \geq 7.5$ km exhibit large concaves, indicating that background seismicity is incorrectly assigned to clusters. For the smallest threshold $\epsilon = 2.5$ km, some small peaks appear and thus the $\epsilon = 5$ km as the optimal value was selected. Dataset D3 contains offshore seismicity in the NAS area, with probably higher location errors. This supports our choice for a larger distance threshold.

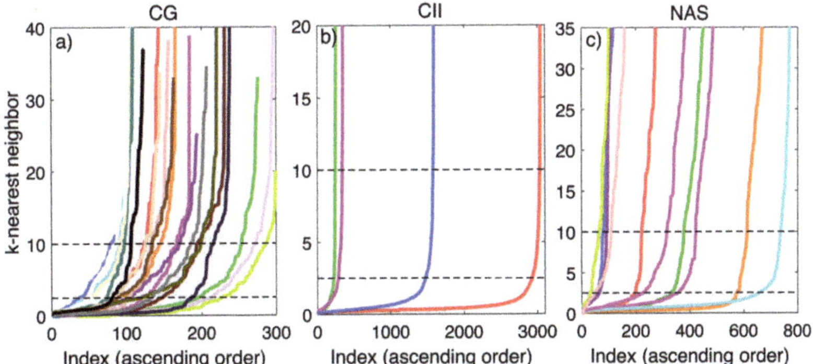

Figure A3. The k-nearest neighbor plot of the potential clusters with $N \geq 100$ events in (**a**) CG (**b**) CII and (**c**) NAS. Black horizontal dashed lines indicate the range of ϵ values given as input to the DBSCAN algorithm and each color corresponds to a potential cluster.

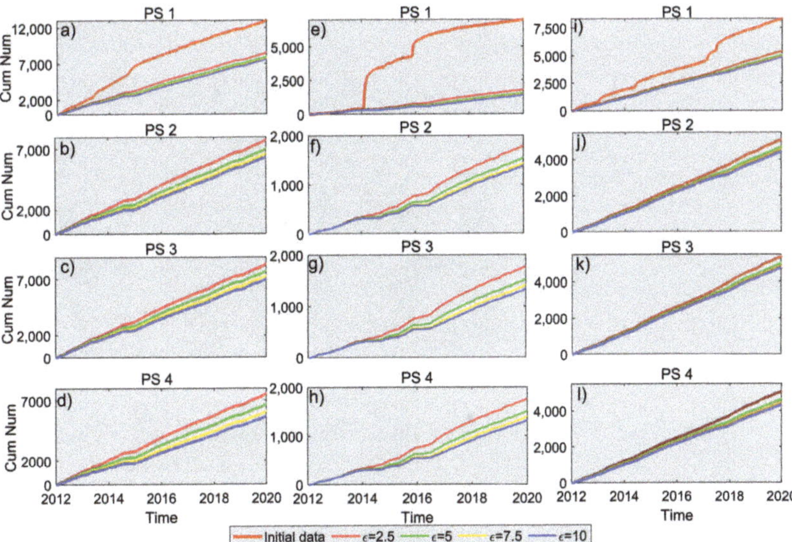

Figure A4. Cumulative number of the initial datasets (red line) and cumulative number of background seismicity for each parameter set (PS1-PS4) and for four different distance thresholds ($\epsilon = 2.5, 5, 7.5, 10$ km). (**a–d**) Dataset D1, (**e–h**) dataset D2 and (**i–l**) dataset D3.

To further explore the differences between the spatio-temporal evolution of the declustered catalogs, the space-time pattern of the background events is examined, comparing the full and the declustered catalogs. In dataset D1, a persistent gap of seismicity appears during the second half of 2014, independently of the chosen temporal constraints, associated with the two large earthquake swarms in that period [77]. Due to the intense seismic activity during 2013–2014 in the western Corinth Gulf [57,65], the classification of seismicity into clusters becomes more complicated, so we have chosen a rather conservative parameter set, *PS3*, with $T = 0$. In this way, we avoid merging distinct clusters that are spatio-temporally close to each other. Figure A5a shows the space-time evolution of the declustered catalog that corresponds to the final parameter set. The main seismic excitations present in Figure A5b are detected, while preserving the patterns of the background seismicity. In dataset D2, the results are quite similar for all the tested temporal constraints, and for this reason, we adopted parameter set $PS4$ with $T = 5$ days, which is a more loose constrain. It is more likely for seismic excitations close in time to be part of the same main shock–aftershock sequence, due to the two major sequences that dominate in the study period. In the initial dataset (Figure A5c), the two major sequences are visible, whereas they are removed after the implementation of the clustering algorithm, while preserving the main patterns of background seismicity (Figure A5d). Finally, for the NAS area, the differences over the temporal constraints seem negligible, therefore, we chose parameter set $PS4$. Figure A5e illustrates a standard scattering of the background seismicity in space without gaps and high-density areas, whereas the main seismic sequences visible in Figure A5f have been identified.

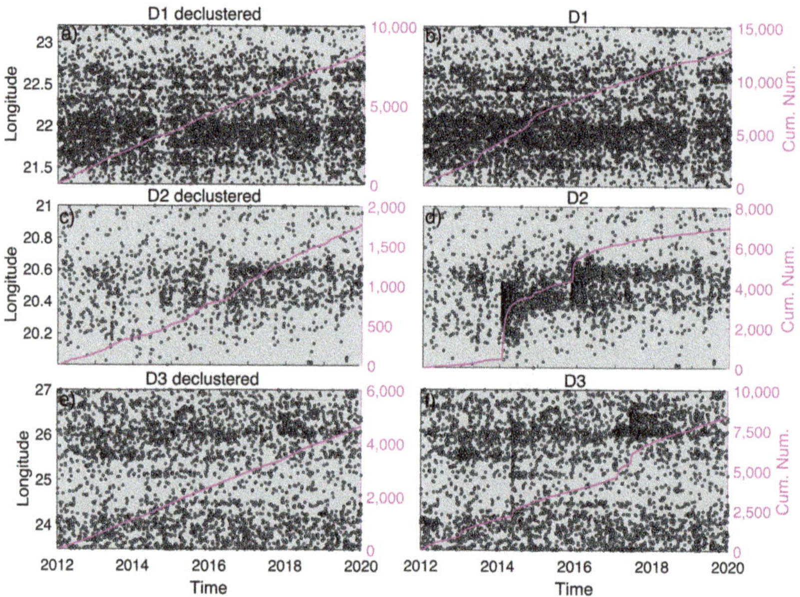

Figure A5. Space-time evolution of the background and initial seismicity for dataset (**a**,**b**) D1, (**c**,**d**) D2 and (**e**,**f**) D3. Purple lines denote the cumulative number of events.

References

1. Ross, Z.E.; Trugman, D.T.; Hauksson, E.; Shearer, P.M. Searching for hidden earthquakes in Southern California. *Science* **2019**, *364*, 767–771. [CrossRef] [PubMed]
2. Omi, T.; Ogata, Y.; Hirata, Y.; Aihara, K. Intermediate-term forecasting of aftershocks from an early aftershock sequence: Bayesian and ensemble forecasting approaches. *J. Geophys. Res.* **2015**, *120*, 2561–2578. [CrossRef]
3. Page, M.T.; Van Der Elst, N.; Hardebeck, J.; Felzer, K.; Michael, A.J. Three ingredients for improved global aftershock forecasts: Tectonic region, time-dependent catalog incompleteness, and intersequence variability. *Bull. Seism. Soc. Am.* **2016**, *106*, 2290–2301. [CrossRef]
4. Petersen, G.; Niemz, P.; Cesca, S.; Mouslopoulou, V.; Bocchini, G. Clusty, the waveform-based network similarity clustering toolbox: concept and application to image complex faulting offshore Zakynthos (Greece). *Geophys. J. Int.* **2021**, *224*, 2044–2059. [CrossRef]
5. Kamer, Y.; Ouillon, G.; Sornette, D. Fault network reconstruction using agglomerative clustering: applications to southern Californian seismicity. *Nat. Hazard Earth Syst.* **2020**, *20*, 3611–3625. [CrossRef]
6. Petersen, M.D.; Mueller, C.S.; Moschetti, M.P.; Hoover, S.M.; Rukstales, K.S.; McNamara, D.E.; Williams, R.A.; Shumway, A.M.; Powers, P.M.; Earle, P.S.; et al. 2018 One-Year Seismic Hazard Forecast for the Central and Eastern United States from Induced and Natural Earthquakes. *Seismol. Res. Lett.* **2018**, *89*, 1049–1061. [CrossRef]

7.   Mizrahi, L.; Nandan, S.; Wiemer, S. The effect of declustering on the size distribution of mainshocks. *Seismol. Res. Lett.* **2021**. [CrossRef]

8.   Taroni, M.; Akinci, A. Good practices in PSHA: Declustering, b-value estimation, foreshocks and aftershocks inclusion; a case study in Italy. *Geophys. J. Int.* **2021**, *224*, 1174–1187. [CrossRef]

9.   Llenos, A.L.; Michael, A.J. Regionally optimized background earthquake rates from ETAS (ROBERE) for probabilistic seismic hazard assessment. *Bull. Seism. Soc. Am.* **2020**, *110*, 1172–1190. [CrossRef]

10.   Gardner, J.; Knopoff, L. Is the sequence of earthquakes in Southern California, with aftershocks removed, Poissonian? *Bull. Seism. Soc. Am.* **1974**, *64*, 1363–1367. [CrossRef]

11.   Peresan, A.; Gentili, S. Identification and characterisation of earthquake clusters: A comparative analysis for selected sequences in Italy and adjacent regions. *Boll. Geofis. Teor. Appl.* **2020**, *61*, 57-80.

12.   Reasenberg, P. Second-order moment of central California seismicity, 1969–1982. *J. Geophys. Res.* **1985**, *90*, 5479–5495. [CrossRef]

13.   Zhuang, J.; Ogata, Y.; Vere-Jones, D. Stochastic declustering of space-time earthquake occurrences. *J. Am. Stat. Assoc.* **2002**, *97*, 369–380. [CrossRef]

14.   Ogata, Y. Statistical models for earthquake occurrences and residual analysis for point processes. *J. Am. Stat. Assoc.* **1988**, *83*, 9–27. [CrossRef]

15.   Ogata, Y. Space-time point-process models for earthquake occurrences. *Ann. I. Stat. Math.* **1998**, *50*, 379–402. [CrossRef]

16.   Zhuang, J.; Ogata, Y.; Vere-Jones, D. Analyzing earthquake clustering features by using stochastic reconstruction. *J. Geophys. Res.* **2004**, *109*, B05301. [CrossRef]

17.   Zhuang, J.; Murru, M.; Falcone, G.; Guo, Y. An extensive study of clustering features of seismicity in Italy from 2005 to 2016. *Geophys. J. Int.* **2019**, *216*, 302–318. [CrossRef]

18.   Zhou, P.; Yang, H.; Wang, B.; Zhuang, J. Seismological investigations of induced earthquakes near the Hutubi underground gas storage facility. *J. Geophys. Res.* **2019**, *124*, 8753–8770. [CrossRef]

19.   Marsan, D.; Prono, E.; Helmstetter, A. Monitoring aseismic forcing in fault zones using earthquake time series. *Bull. Seismol. Soc. Am.* **2013**, *103*, 169–179. [CrossRef]

20.   Crespo-Martín, C.; Martín-González, F.; Yazdi, P.; Hainzl, S.; Rincón, M. Time-dependent and spatiotemporal statistical analysis of intraplate anomalous seismicity: Sarria-Triacastela-Becerreá (NW Iberian Peninsula, Spain). *Geophys. J. Int.* **2021**, *225*, 477–493. [CrossRef]

21.   Peng, W.; Marsan, D.; Chen, K.H.; Pathier, E. Earthquake swarms in Taiwan: A composite declustering method for detection and their spatial characteristics. *Earth Planet. Sci. Lett.* **2021**, *574*, 117160. [CrossRef]

22.   Baiesi, M.; Paczuski, M. Scale-free networks of earthquakes and aftershocks. *Phys. Rev. E* **2004**, *69*, 066106. [CrossRef] [PubMed]

23.   Zaliapin, I.; Ben-Zion, Y. Earthquake clusters in southern California I: Identification and stability. *J. Geophys. Res.* **2013**, *118*, 2847–2864. [CrossRef]

24.   Zaliapin, I.; Ben-Zion, Y. A global classification and characterization of earthquake clusters. *Geophys. J. Int.* **2016**, *207*, 608–634. [CrossRef]

25.   Peresan, A.; Gentili, S. Seismic clusters analysis in Northeastern Italy by the nearest-neighbor approach. *Phys. Earth Planet Inter.* **2018**, *274*, 87–104. [CrossRef]

26.   Martínez-Garzón, P.; Ben-Zion, Y.; Zaliapin, I.; Bohnhoff, M. Seismic clustering in the Sea of Marmara: Implications for monitoring earthquake processes. *Tectonophysics* **2019**, *768*, 228176. [CrossRef]

27.   Bayliss, K.; Naylor, M.; Main, I.G. Probabilistic identification of earthquake clusters using rescaled nearest neighbour distance networks. *Geophys. J. Int.* **2019**, *217*, 487–503. [CrossRef]

28.   Bottiglieri, M.; Lippiello, E.; Godano, C.; De Arcangelis, L. Identification and spatiotemporal organization of aftershocks. *J. Geophys. Res.* **2009**, *114*, B03303. [CrossRef]

29.   Jacobs, K.M.; Smith, E.G.; Savage, M.K.; Zhuang, J. Cumulative rate analysis (CURATE): A clustering algorithm for swarm dominated catalogs. *J. Geophys. Res.* **2013**, *118*, 553–569. [CrossRef]

30.   Neuts, M.F. A Versatile Markovian Point Process. *J. Appl. Probab.* **1979**, *16*, 764–779. [CrossRef]

31.   Bountzis, P.; Papadimitriou, E.; Tsaklidis, G. Earthquake clusters identification through a Markovian Arrival Process (MAP): Application in Corinth Gulf (Greece). *Physica A* **2020**, *545*, 123655. [CrossRef]

32.   Lu, S. A Bayesian multiple changepoint model for marked poisson processes with applications to deep earthquakes. *Stoch. Environ. Res. Risk A* **2019**, *33*, 59–72.

33.   Benali, A.; Peresan, A.; Varini, E.; Talbi, A. Modelling background seismicity components identified by nearest neighbour and stochastic declustering approaches: The case of Northeastern Italy. *Stoch. Environ. Res. Risk A* **2020**, *34*, 775–791. [CrossRef]

34.   Bountzis, P.; Kostoglou, A.; Papadimitriou, E.; Karakostas, V. Identification of spatiotemporal seismicity clusters in central Ionian Islands (Greece). *Phys. Earth Planet. Inter.* **2021**, *312*, 106675. [CrossRef]

35.   Ester, M.; Kriegel, H.P.; Sander, J.; Xu, X. A density-based algorithm for discovering clusters in large spatial databases with noise. In Proceedings of the Second International Conference on Knowledge Discovery and Data Mining, Portland, OR, USA, 2–4 August 1996; Volume 96, pp. 226–231.

36.   Llenos, A.L.; Michael, A.J. Forecasting the (un) productivity of the 2014 M 6.0 South Napa aftershock sequence. *Seismol. Res. Lett.* **2017**, *88*, 1241–1251. [CrossRef]

37. Hardebeck, J.L.; Llenos, A.L.; Michael, A.J.; Page, M.T.; Van Der Elst, N. Updated California aftershock parameters. *Seismol. Res. Lett.* **2019**, *90*, 262–270. [CrossRef]

38. Utsu, T.; Ogata, Y. The centenary of the Omori formula for a decay law of aftershock activity. *J. Phys. Earth* **1995**, *43*, 1–33. [CrossRef]

39. Hainzl, S.; Ogata, Y. Detecting fluid signals in seismicity data through statistical earthquake modeling. *J. Geophys. Res.* **2005**, *110*, B05S07. [CrossRef]

40. Marsan, D.; Reverso, T.; Helmstetter, A.; Enescu, B. Slow slip and aseismic deformation episodes associated with the subducting Pacific plate offshore Japan, revealed by changes in seismicity. *J. Geophys. Res.* **2013**, *118*, 4900–4909. [CrossRef]

41. Crespo Martín, C.; Martín-González, F. Statistical Analysis of Intraplate Seismic Clusters: The Case of the NW Iberian Peninsula. *Pure Appl. Gephys.* **2021**, *178*, 3355–3374. [CrossRef]

42. Lippiello, E.; Godano, C.; de Arcangelis, L. The Relevance of Foreshocks in Earthquake Triggering: A Statistical Study. *Entropy* **2019**, *21*, 173. [CrossRef] [PubMed]

43. Cesca, S. Seiscloud, a tool for density-based seismicity clustering and visualization. *J. Seismol.* **2020**, *24*, 443–457. [CrossRef]

44. Cesca, S.; Grigoli, F.; Heimann, S.; Dahm, T.; Kriegerowski, M.; Sobiesiak, M.; Tassara, C.; Olcay, M. The Mw 8.1 2014 Iquique, Chile, seismic sequence: A tale of foreshocks and aftershocks. *Geophys. J. Int.* **2016**, *204*, 1766–1780. [CrossRef]

45. Sheikhhosseini, Z.; Mirzaei, N.; Heidari, R.; Monkaresi, H. Delineation of potential seismic sources using weighted K-means cluster analysis and particle swarm optimization (PSO). *Acta Geophys.* **2021**, *69*, 2161–2172. [CrossRef]

46. Fortunato, S.; Hric, D. Community detection in networks: A user guide. *Phys. Rep.* **2016**, *659*, 1–44. [CrossRef]

47. Lippiello, E.; Bountzis, P. An objective criterion for cluster detection in stochastic epidemic models. *arXiv* **2021**, arXiv:2104.04138.

48. Hatzfeld, D.; Karakostas, V.; Ziazia, M.; Kassaras, I.; Papadimitriou, E.; Makropoulos, K.; Voulgaris, N.; Papaioannou, C. Microseismicity and faulting geometry in the Gulf of Corinth (Greece). *Geophys. J. Int.* **2000**, *141*, 438–456. [CrossRef]

49. Scordilis, E.; Karakaisis, G.; Karacostas, B.; Panagiotopoulos, D.; Comninakis, P.; Papazachos, B. Evidence for transform faulting in the Ionian Sea: The Cephalonia island earthquake sequence of 1983. *Pure Appl. Gephys.* **1985**, *123*, 388–397. [CrossRef]

50. Louvari, E.; Kiratzi, A.; Papazachos, B. The Cephalonia transform fault and its extension to western Lefkada Island (Greece). *Tectonophysics* **1999**, *308*, 223–236. [CrossRef]

51. Papazachos, B.; Papadimitriou, E.; Kiratzi, A.; Papazachos, C.; Louvari, E. Fault plane solutions in the Aegean Sea and the surrounding area and their tectonic implication. *Boll. Geof. Teor. Appl.* **1998**, *39*, 199–218.

52. McKenzie, D. Active tectonics of the Mediterranean region. *Geophys. J. Int.* **1972**, *30*, 109–185. [CrossRef]

53. Le Pichon, X.; Angelier, J. The Hellenic arc and trench system: A key to the neotectonic evolution of the eastern Mediterranean area. *Tectonophysics* **1979**, *60*, 1–42. [CrossRef]

54. Permanent Regional Seismological Network. (Operated by the Aristotle University of Thessaloniki). International Federation of Digital Seismograph Networks. 1981. Available online: http://dx.doi.org/10.7914/SN/HT (accessed on 15 January 2021).

55. Wiemer, S.; Wyss, M. Minimum magnitude of completeness in earthquake catalogs: Examples from Alaska, the western United States, and Japan. *Bull. Seismol. Soc. Am.* **2000**, *90*, 859–869. [CrossRef]

56. Aki, K. Maximum likelihood estimate of b in the formula logN= a-bM and its confidence limits. *Bull. Earthq. Res. Inst. Tokyo Univ.* **1965**, *43*, 237–239.

57. Kapetanidis, V.; Michas, G.; Kaviris, G.; Vallianatos, F. Spatiotemporal Properties of Seismicity and Variations of Shear-Wave Splitting Parameters in the Western Gulf of Corinth (Greece). *Appl. Sci.* **2021**, *11*, 6573. [CrossRef]

58. Karakostas, V.; Papadimitriou, E.; Mesimeri, M.; Gkarlaouni, C.; Paradisopoulou, P. The 2014 Kefalonia doublet (Mw 6.1 and Mw 6.0), central Ionian Islands, Greece: Seismotectonic implications along the Kefalonia transform fault zone. *Acta Geophys.* **2015**, *63*, 1–16. [CrossRef]

59. Papadimitriou, E.; Karakostas, V.; Mesimeri, M.; Chouliaras, G.; Kourouklas, C. The Mw6. 5 17 November 2015 Lefkada (Greece) earthquake: Structural interpretation by means of the aftershock analysis. *Pure Appl. Gephys.* **2017**, *174*, 3869–3888. [CrossRef]

60. Karakostas, V.; Papadimitriou, E.; Gospodinov, D. Modelling the 2013 North Aegean (Greece) seismic sequence: geometrical and frictional constraints, and aftershock probabilities. *Geophys. J. Int.* **2014**, *197*, 525–541. [CrossRef]

61. Saltogianni, V.; Gianniou, M.; Taymaz, T.; Yolsal-Çevikbilen, S.; Stiros, S. Fault slip source models for the 2014 Mw 6.9 Samothraki-Gökçeada earthquake (North Aegean trough) combining geodetic and seismological observations. *J. Geophys. Res.* **2015**, *120*, 8610–8622. [CrossRef]

62. Papadimitriou, P.; Kassaras, I.; Kaviris, G.; Tselentis, G.A.; Voulgaris, N.; Lekkas, E.; Chouliaras, G.; Evangelidis, C.; Pavlou, K.; Kapetanidis, V.; et al. The 12th June 2017 Mw = 6.3 Lesvos earthquake from detailed seismological observations. *J. Geodyn.* **2018**, *115*, 23–42. [CrossRef]

63. Kapetanidis, V.; Deschamps, A.; Papadimitriou, P.; Matrullo, E.; Karakonstantis, A.; Bozionelos, G.; Kaviris, G.; Serpetsidaki, A.; Lyon-Caen, H.; Voulgaris, N.; et al. The 2013 earthquake swarm in Helike, Greece: Seismic activity at the root of old normal faults. *Geophys. J. Int.* **2015**, *202*, 2044–2073. [CrossRef]

64. Mesimeri, M.; Karakostas, V.; Papadimitriou, E.; Schaff, D.; Tsaklidis, G. Spatio-temporal properties and evolution of the 2013 Aigion earthquake swarm (Corinth Gulf, Greece). *J. Seismol.* **2016**, *20*, 595–614. [CrossRef]

65. Michas, G.; Kapetanidis, V.; Kaviris, G.; Vallianatos, F. Earthquake Diffusion Variations in the Western Gulf of Corinth (Greece). *Pure Appl. Gephys.* **2021**, *178*, 2855–2870. [CrossRef]

66. Kapetanidis, V. Spatiotemporal Patterns of Microseismicity for the Identification of Active Fault Structures Using Seismic Waveform Cross-Correlation and Double-Difference Relocation. Ph.D. Thesis, Department of Geophysics-Geothermics, Faculty of Geology and Geoenvironment, University of Athens, Athens, Greece, 2017.
67. Mesimeri, M.; Kourouklas, C.; Papadimitriou, E.; Karakostas, V.; Kementzetzidou, D. Analysis of microseismicity associated with the 2017 seismic swarm near the Aegean coast of NW Turkey. *Acta Geophys.* **2018**, *66*, 479–495. [CrossRef]
68. Mesimeri, M.; Karakostas, V.; Papadimitriou, E.; Tsaklidis, G. Characteristics of earthquake clusters: Application to western Corinth Gulf (Greece). *Tectonophysics* **2019**, *767*, 228160. [CrossRef]
69. Hainzl, S.; Zakharova, O.; Marsan, D. Impact of aseismic transients on the estimation of aftershock productivity parameters. *Bull. Seism. Soc. Am.* **2013**, *103*, 1723–1732. [CrossRef]
70. Llenos, A.L.; McGuire, J.J.; Ogata, Y. Modeling seismic swarms triggered by aseismic transients. *Earth Planet. Sci. Lett.* **2009**, *281*, 59–69. [CrossRef]
71. Mesimeri, M.; Karakostas, V. Repeating earthquakes in western Corinth Gulf (Greece): Implications for aseismic slip near locked faults. *Geophys. J. Int.* **2018**, *215*, 659–676. [CrossRef]
72. Wessel, P.; Smith, W.H.; Scharroo, R.; Luis, J.; Wobbe, F. Generic mapping tools: Improved version released. *Eos Trans. Am. Geophys. Union* **2013**, *94*, 409–410. [CrossRef]
73. Dempster, A.P.; Laird, N.M.; Rubin, D.B. Maximum likelihood from incomplete data via the EM algorithm. *J. R. Stat. Soc. Ser. B Stat. Methodol.* **1977**, *39*, 1–38.
74. Schwarz, G. Estimating the dimension of a model. *Ann. Stat.* **1978**, *6*, 461–464. [CrossRef]
75. van Stiphout, T.; Zhuang, J.; Marsan, D. Seismicity declustering. *Community Online Resour. Stat. Seism. Anal.* **2012**, *10*, 1.
76. Wiemer, S. A software package to analyze seismicity: ZMAP. *Seismol. Res. Lett.* **2001**, *72*, 373–382. [CrossRef]
77. Duverger, C.; Lambotte, S.; Bernard, P.; Lyon-Caen, H.; Deschamps, A.; Nercessian, A. Dynamics of microseismicity and its relationship with the active structures in the western Corinth Rift (Greece). *Geophys. J. Int.* **2018**, *215*, 196–221. [CrossRef]

Article

A New Smoothed Seismicity Approach to Include Aftershocks and Foreshocks in Spatial Earthquake Forecasting: Application to the Global $M_w \geq 5.5$ Seismicity

Matteo Taroni * and Aybige Akinci

Istituto Nazionale di Geofisica e Vulcanologia, INGV Sez. Roma 1 Sismologia e Tettonofisica, Via di Vigna Murata 605, 00143 Roma, Italy; aybige.akinci@ingv.it
* Correspondence: matteo.taroni@ingv.it

Abstract: Seismicity-based earthquake forecasting models have been primarily studied and developed over the past twenty years. These models mainly rely on seismicity catalogs as their data source and provide forecasts in time, space, and magnitude in a quantifiable manner. In this study, we presented a technique to better determine future earthquakes in space based on spatially smoothed seismicity. The improvement's main objective is to use foreshock and aftershock events together with their mainshocks. Time-independent earthquake forecast models are often developed using declustered catalogs, where smaller-magnitude events regarding their mainshocks are removed from the catalog. Declustered catalogs are required in the probabilistic seismic hazard analysis (PSHA) to hold the Poisson assumption that the events are independent in time and space. However, as highlighted and presented by many recent studies, removing such events from seismic catalogs may lead to underestimating seismicity rates and, consequently, the final seismic hazard in terms of ground shaking. Our study also demonstrated that considering the complete catalog may improve future earthquakes' spatial forecast. To do so, we adopted two different smoothed seismicity methods: (1) the fixed smoothing method, which uses spatially uniform smoothing parameters, and (2) the adaptive smoothing method, which relates an individual smoothing distance for each earthquake. The smoothed seismicity models are constructed by using the global earthquake catalog with $M_w \geq 5.5$ events. We reported progress on comparing smoothed seismicity models developed by calculating and evaluating the joint log-likelihoods. Our resulting forecast shows a significant information gain concerning both fixed and adaptive smoothing model forecasts. Our findings indicate that complete catalogs are a notable feature for increasing the spatial variation skill of seismicity forecasts.

Keywords: smoothed seismicity methods; global seismicity; foreshocks and aftershocks; earthquake forecasting model

Citation: Taroni, M.; Akinci, A. A New Smoothed Seismicity Approach to Include Aftershocks and Foreshocks in Spatial Earthquake Forecasting: Application to the Global $M_w \geq 5.5$ Seismicity. *Appl. Sci.* **2021**, *11*, 10899. https://doi.org/10.3390/app112210899

Academic Editor: Stefania Gentili

Received: 13 September 2021
Accepted: 8 November 2021
Published: 18 November 2021

Publisher's Note: MDPI stays neutral with regard to jurisdictional claims in published maps and institutional affiliations.

1. Introduction

Building earthquake forecasting models is a fundamental step in any probabilistic seismic hazard analysis (PSHA). The spatial distribution of future seismicity is usually estimated using a seismicity catalog using two commonly adopted approaches called zonation [1,2] and smoothed seismicity [3,4]. In this work, we focus our attention on the smoothed seismicity approach. This approach uses statistical techniques to build a spatially gridded model using the epicenters of seismic events. One of the first examples of the smoothed seismicity model was developed by [3] and used the Gaussian isotropic spatial kernel to smooth the seismicity around epicenters. This model is based on only one parameter, i.e., the sigma of the Gaussian kernel: the larger the sigma, the larger the smoothing and vice versa. In the Frankel model, the sigma is fixed for any event, so it is called "fixed smoothed seismicity". Later, [4] developed a smoothed seismicity model that allows changing the sigma of the Gaussian kernel, and in general the size of any spatial kernel function, according to the local density of earthquakes. The idea of this model is that where we have more events, we can use a smaller sigma to better define the seismic structures (i.e., the faults) that generate the seismicity. On the other hand, where we have fewer events, we can use a larger sigma to increase the coverage of the model in those lower seismogenic zones. In traditional PSHA, earthquakes are modeled using a Poisson process, where the occurrence of a future earthquake is independent of previous earthquakes from the same source [5]. The Poisson hypothesis holds for declustered catalogs. To include aftershocks and foreshocks within traditional PSHA, Ref. [6] presented an approach based on [7] theorem and its consequent generalization [8]. They demonstrated that the Poisson distribution could approximate the distribution of exceedances (also considering seismic sequences) in some specific conditions, e.g., for a probability of 10 percent or less of having an exceedance in 50 years (a typical value used for PSHA). Ref. [9] somewhat revised the initial [6] procedure. Rather than using their correction factor, Ref. [9] employed the b-value and the annual rate of the complete catalog as input for PSHA computations.

Both [6] and [9] suggest using a declustered seismic catalog only for the spatial estimation to avoid spatial bias introduced by the seismic sequence.

Therefore, a method that wants to introduce such sequences in the spatial estimation for PSHA needs a technique to downweigh the importance of aftershocks and foreshocks. Indeed, any seismic sequence should have the same importance in the spatial estimation of seismicity, independently from the number of events in the sequence (which can greatly vary between the sequences). The delcustering technique is the most dichotomous approach: it gives a weight equal to 1 to the mainshock and 0 to all other events in the sequence.

In their pioneering work, Ref. [10] developed a model to determine the spatial distribution of seismicity, including also the aftershocks and foreshocks in the seismic catalog. This approach uses a statistical model for the seismicity triggering, the ETAS model [11] and the stochastic declustering procedure [12] to assign each event the probability to be an independent event. In fact, in the ETAS model, events in the catalogs are distinguished as independent and dependent instead of mainshocks and aftershocks. The aftershocks of a seismic sequence, dependent on the sequence's mainshock, obtain a very low weight in this framework. Ref. [10] model consists of the multiplication of each spatial kernel for the probability to be independent of the associated earthquake. Therefore, in this framework, the spatial density distribution of a seismic sequence is mainly concentrated near the mainshock of the sequence (i.e., the independent event that generates all the dependent events of the sequence). Using this method, the fault that caused the seismic sequence is only partially reconstructed.

Our new, simple approach tries to solve that problem using a uniform weight for all the events of the same seismic sequence (i.e., 1/M, where M is the number of events in the seismic sequence). In this manner, it is possible to describe the fault or the system of faults in a more coherent way, avoiding giving excessive weight to the mainshock of the sequence. Here, we use the global seismic catalog (CMT catalog), Ref. [13] to build four different spatial seismicity models, fixed and adaptive smoothed seismicity with and without our correction, to take into account the seismic sequences. Finally, we use the last ten years of the catalog to compare the performances of the models, using the spatial likelihoods of the models to measure their efficiency.

2. Dataset

We used the global centroid moment tensor (CMT) catalog containing 11,638 earthquakes with a depth ≤ 50 km recorded over the past almost 40 years between 1980 and 2019 [13,14]. We considered only events above the completeness magnitude as threshold $M_w = 5.5$ [13,15]. The epicenter distribution of these events is shown in Figure 1. The current seismic sequences present in the seismic catalog have been detected by the [16] declustering algorithm, and the related parameters are provided and implemented in the ZMAP software [17]. Figure 2 shows the mainshocks (red dots) and foreshocks/aftershocks (blue dots) in three zones in the world (Chile, Mexico, and Indonesia). Table 1 shows the number of events present in each subcatalog. We stress that the declustered catalog (i.e., the catalog containing only the mainshocks of the sequences) has 6440 events, about 45% less with respect to the complete catalog. This work aims to maintain as much data as possible and use all the available earthquakes in the catalog for the spatial distribution modeling. We underline that the use of a global catalog, instead of regional catalogs, has some drawbacks: a high threshold for the completeness magnitude (in our case M_w 5.5), difficulty in recognizing volcanic events, and large uncertainties in hypocentral estimation. The main advantage is a large number of strong events, which can be collected in a few years.

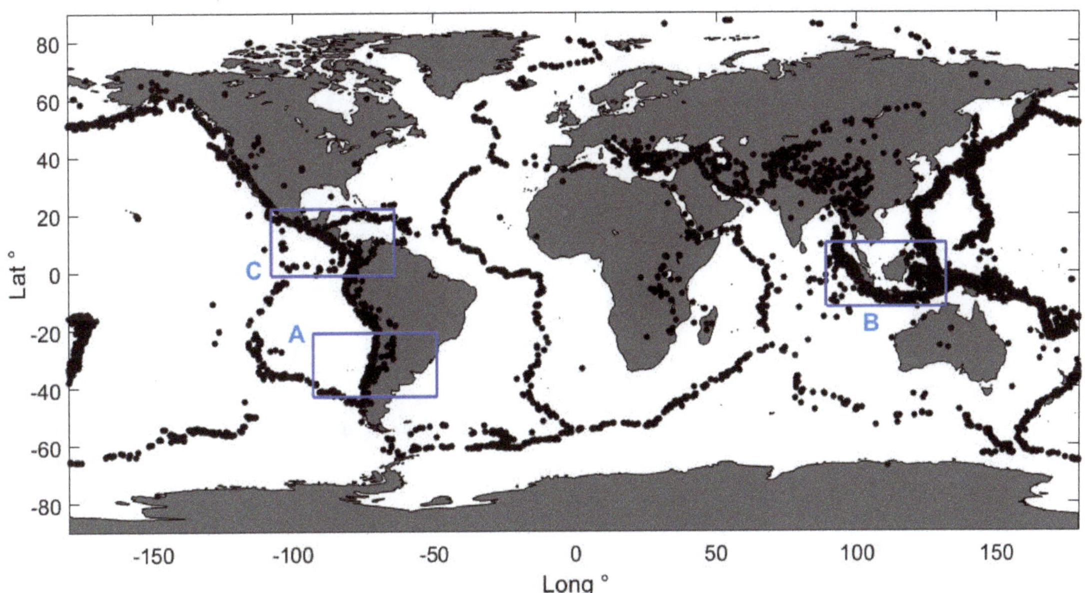

Figure 1. Location of earthquakes in the global centroid moment tensor (CMT) catalog with a depth $\leq$ 50 km recorded over the past almost 40 years between 1980 and 2019 [13,14]; blue letters indicate the zones of the zoom-in Figure 2.

Table 1. Number of events and time windows in the different catalogs from $M_w \geq 5.5$.

Catalog Type	Time Window	Number of Events
Complete	1980–2019	11638
Declustered	1980–2019	6440
Complete–Learning	1980–2009	7977
Declustered–Learning	1980–2009	4718
Complete–Testing	2010–2019	3161

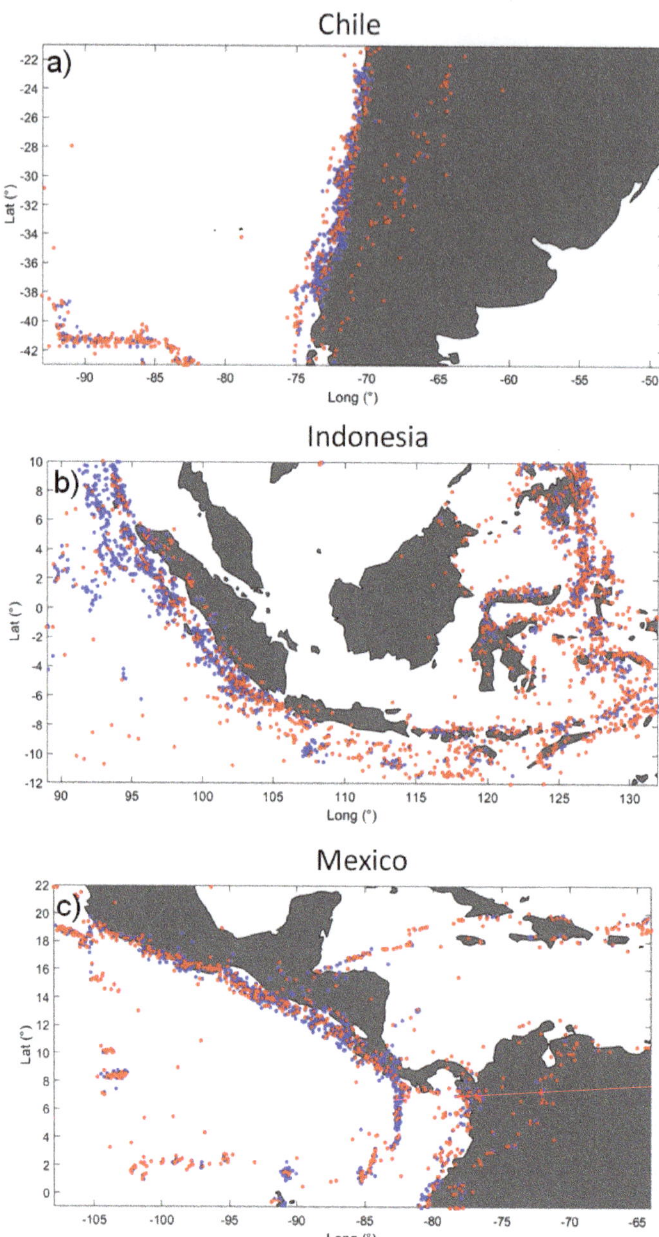

Figure 2. Location of earthquakes in the global centroid moment tensor (CMT) catalog with a depth ≤ 50 km recorded over the past almost 40 years between 1980 and 2019 [13,14]; (**a–c**) show the mainshocks (red dots) and foreshocks/aftershocks (blue dots) in some zones in the world: Indonesia, Mexico, and Chile.

3. A New Smoothed Seismicity Approach

Building a spatial grid is the first step to constructing a spatial smoothed seismicity model [3,4]. In this work, we used a global spatial regular grid, 0.5° by 0.5°. Therefore, we need to compute the contribution to each event in the seismic catalog to the generic *i*-th spatial grid; the following equation describes that contribution:

$$f_i = \sum_{j=1}^{N} cK_{ij}A_i d_j \tag{1}$$

where f_i represents the normalized total seismic rate for the *i*-th spatial grid, N is the total number of events in the complete (i.e., not declustered) catalog, c is the normalization factor $\left(c = \frac{1}{\sum f_i}\right)$, K_{ij} is the kernel function that depends on the distance between the center of the *i-th* spatial cell and the epicentre of the *j*-th earthquake, A_i is the area of the *i-th* spatial cell, and d_j is the correction to take into account the foreshocks and aftershocks contribution to the spatial model.

The following Gaussian kernel function [3] is used:

$$K_{ij} = \frac{1}{2\pi\sigma^2} e^{-\frac{r_{ij}^2}{2\sigma^2}} \tag{2}$$

where r_{ij} is the distance between the center of the *i*-th spatial cell and the epicentre of the *j*-th earthquake, and σ is the free parameter of the model that rules the amplitude of the smoothing. However, we note that different kernel functions can also be employed in smoothing the epicenters from the earthquake catalog [4,18]. The smoothing distance, σ, involved in each earthquake may be defined differently in various smoothed seismicity models. For example, the fixed smoothed seismicity models practiced a single smoothing distance for all earthquakes. The adaptive smoothed seismicity models represent unique smoothing distances for each earthquake between an event and its *n*th closest neighbors (NN), resulting in spatially varying smoothing distances [4]. The distance becomes smaller in regions of high seismicity than in areas with sparse seismicity. It is one of the crucial parameters in the smoothed seismicity models both for the earthquake rates and the spatial variations of the earthquake activity rates in a region [19]. The correction parameter d_j represents the innovative part of our method. It is defined as following $d_j = \frac{1}{S_j}$, where S_j is the number of events in the seismic sequence and contains the *j*-th event. For example, if a seismic sequence contains ten events, one mainshock, and nine aftershocks, each event receives a weight, $= \frac{1}{10}$. Since the sum of all the weights is equal to one, the inclusion of aftershocks does not create a spatial bias in the model [6], and it leads to a better description of the fault that generated the sequence.

Using this simple correction may help better identify the active fault structures and their features in a region. Removing all the aftershocks and foreshocks [3,4], giving very high weight to the mainshocks only [10], may lead to an incomplete or biased view of the spatial distribution of future seismicity. Conversely, considering all the events in the sequence with a uniform weight, as in our method, increases the model's forecasting performance.

We underline that with Equation (1), we build normalized smoothed seismicity models, i.e., the sum of all the rates in the spatial cells are equal to 1. In this work, we do not face the problem of the total number of events and their magnitude frequency distribution, already treated in [9]. In that work, the seismicity rates are corrected by a proposed technique that allows counting all events in the complete seismic catalog by quickly adjusting the magnitude frequency relationships. Our method differs from theirs, since we only deal with the spatial distribution of the seismicity by using an equal weight for all the events of the corresponding seismic sequence and incorporating aftershocks to improve the spatial resolution of the model.

4. Likelihood Testing for Spatial Variation of Seismicity

To perform the maximum likelihood estimation of the parameter σ (both for the fixed and adaptive smoothing approach) and to assess the performance of the model, we avoid considering the Poisson distribution of seismic events because this assumption is rarely satisfied by the seismic catalogs [20,21]. Since we are interested only in the spatial distribution of the events, and with Equation (1), we model the normalized spatial distribution of events, we defined the log-likelihood (LL) of the observations with:

$$LL(X|M) = \sum_{i=1}^{N} log(f_i) \tag{3}$$

where X is the set of the N observations (i.e., the epicenters of the events in the seismic catalog), M is the spatial model, log is the natural logarithm, and f_i is the seismic rate of the spatial cell where the i-th event is located. We note that this formulation differs from the spatial LL defined by [22] and has been commonly used in many seismic experiments [23], since the Poisson hypothesis has been abandoned in our study. The LL of Equation (3) may be ratified as the classical LL of a bivariate probability density function (represented by the model, M) in case we assume the independence between the observations in the set X. Additionally, in the case of nonindependent observations, the LL can be still used for scoring the models (some authors, in this case, called the function "pseudo-likelihood", [24].

To perform a pseudoprospective evaluation of the models first, we calculated the log-likelihood values by dividing the earthquake catalog into two parts: (1) the learning catalog, which contains the events recorded between 1980 and 2009 and is used to construct trial smoothed seismicity models, and (2) the testing catalog, which covers the last ten years of catalogs (2010–2019). The same LL of Equation (3) is also used to evaluate the performance of the models.

We applied the fixed and adaptive smoothing methods with and without our correction to include aftershocks and foreshocks for a total number of four different models. First, we used the learning catalog to compute the optimal smoothing parameters from the maximum-likelihood estimations (MLE), which strongly vary with smoothing distance (fixed smoothing) and neighbor number (adaptive smoothing). In the case of fixed smoothing, we used a vector of possible sigma (from 5 km to 200 km, with a spacing of 5 km), while for the adaptive smoothing, a set of possible neighbor numbers) are considered from 1 to 20, with a spacing of 1. The first part of the learning catalog (1980–1999) with a period of twenty years is utilized to build various smoothed seismicity models with different sigma and NN values. Finally, the nearest neighbor numbers and the correlation distances are calculated through maximum-likelihood optimization for the four smoothed seismicity models using the last ten years of the learning catalog (2000–2009). The results of these estimations are summarized in Table 2 and Figure 3.

We underline that these obtained MLE values are suitable only in the case of a global catalog: regional estimation of these parameters can lead to different MLE values (e.g., smaller sigma and larger NN).

Table 2. MLE of the parameters.

Model	MLE
Fixed	Sigma = 135
Adaptive	NN = 1
Corrected Fixed	Sigma = 115
Corrected Adaptive	NN = 1

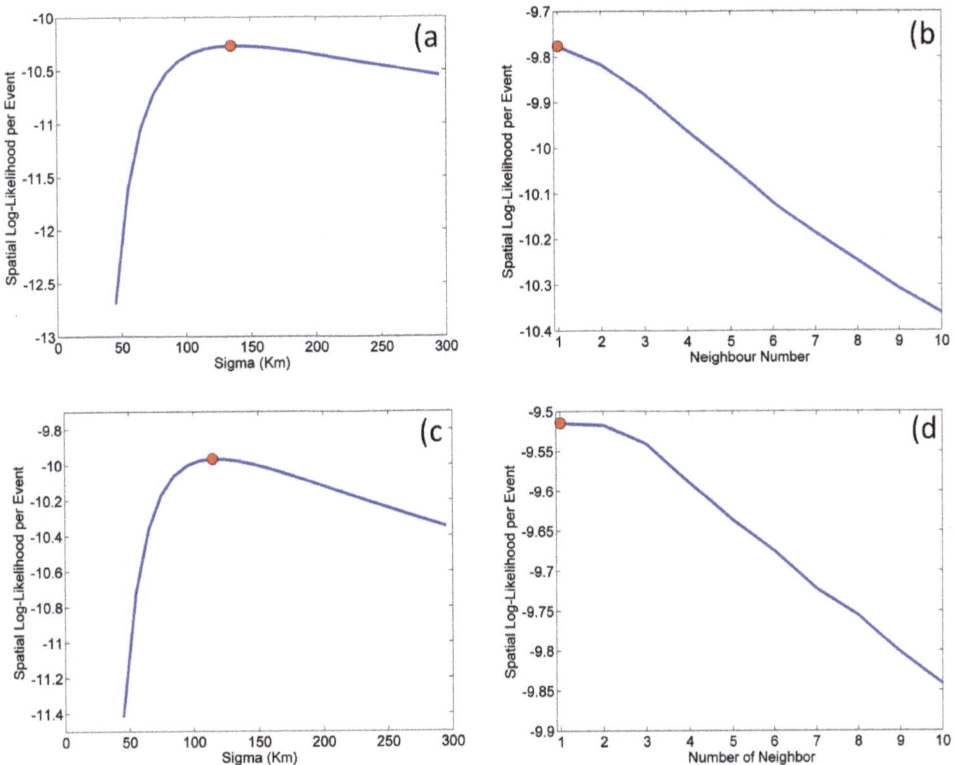

Figure 3. MLE for the parameters sigma and NN for the Fixed (**a**), Adaptive (**b**), Fixed$_{corrected}$ (**c**), and Adaptive$_{corrected}$ (**d**) models. Blue curves represent the log-likelihood functions, red dots the maximum of these functions.

5. Results

The final smoothed seismicity models are constructed using the entire learning catalog and the optimized correlation distances, previously obtained and given in Table 2. The models represent the bidimensional probability density function (PDF) of the seismicity (the sum of all the rates is 1). The corrected fixed smoothed seismicity model is calculated with a smoothing distance of 115 km, and it is 135 km in the case of the uncorrected model. Both adaptive smoothed seismicity models are determined using the nearest neighbor number equal to 1. These fixed and adaptive smoothed seismicity rate models are illustrated in Figure 4a,b (not corrected, hereafter fixed and adaptive) and Figure 4c,d (corrected, hereafter Fixed$_{corrected}$ and Adaptive$_{corrected}$).

To check if our corrected models perform better than those uncorrected smoothed seismicity models, we tested the Fixed$_{corrected}$ and Adaptive$_{corrected}$ models against the two standard fixed and adaptive smoothed seismicity models. Therefore, we performed a global pseudoprospective test, computing the *LL* (Equation (2)) of the four models using the ten-year testing catalog (2010–2019). Here, we outline that our testing catalog is entirely independent of the developed models. We preferred to endorse a similar computation procedure adopted in the real global prospective tests of the Collaboratory for the Study of Earthquake Predictability, CSEP, [23] and the global experiments [25,26]. We evaluated the performance of the models using two different magnitude thresholds, M$_w$ 5.5+ and M$_w$6.5+, to check the robustness of our models' forecasting locations and rates for future earthquakes. The results of these comparisons are presented in Tables 3 and 4 for the four developed models.

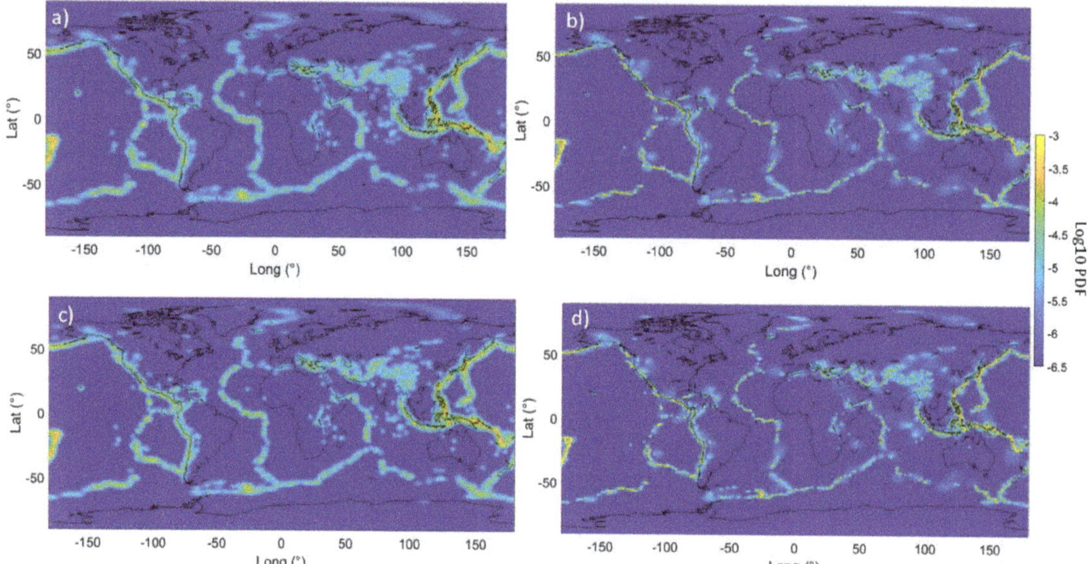

Figure 4. Spatially smoothed seismicity models using (**a**) 135 km smoothing distance from the fixed and (**b**) the nearest neighbor number NN = 1 from the adaptive smoothing seismicity approaches; spatially smoothed seismicity-corrected models using (**c**) 115 km smoothing distance from the fixed and (**d**) the nearest neighbor number NN = 1 from the adaptive smoothing seismicity approaches, employing the epicenters of the earthquakes for $M_w \geq 5.5$ in the global CMT catalog (normalized seismicity rates, i.e., PDF, are in log10 scale).

Table 3. Log-likelihood (*LL*) values for the smoothed seismicity models for testing catalog from magnitude M_w 5.5 (3161 events).

Model	Log-Likelihood (*LL*)
Corrected Adaptive	−29,632
Adaptive	−29,639
Corrected Fixed	−31,198
Fixed	−31,297

Table 4. Log-likelihood (*LL*) values for the smoothed seismicity models for testing catalog from magnitude M_w 6.5 (300 events).

Model	Log-Likelihood (*LL*)
Corrected Adaptive	−2850
Adaptive	−2857
Corrected Fixed	−2931
Fixed	−2949

For a correct interpretation of the models' *LL*, we recall that large *LL* values (i.e., the ones nearest to zero) indicate relatively good performances of the models, and small *LL* values (i.e., the ones further from zero) indicate relative bad performances of the models.

In general, our results show that the adaptive smoothed seismicity models (Adoptive and Adaptive$_{corrected}$) produce larger *LL* values and reveal better forecasting performance with respect to those from the fixed smoothed ones (Fixed and Fixed$_{corrected}$). The *LL* values are −29,632 and −29,639 for the corrected and uncorrected adaptive smoothed models, while they are −31,297 and −31,198 in the case of the fixed corrected and corrected smoothed seismicity models, respectively (Table 3). The largest *LL* value calculated for the adopted smoothed seismicity models arises from the use of the correction parameter including the foreshocks and aftershocks in the global catalog. So, in general, including smaller earthquakes in the clusters increases the performance of the future $M_W \geq 5.5$ and $M_W \geq 6.5$ earthquake forecasting capability in the smoothed seismicity models.

To understand if this increase is rather significant, we interpreted the difference of the *LL* values for two models in terms of the Bayes factor [27], a common interpretation for pseudoprospective experiments [28–30]. According to [27] table, we obtained "very strong evidence" (difference in log-likelihood $\Delta LL > 5$) in favor of our proposed method, both for the fixed and adaptive approaches (Table 5). In Figure 5a–c, we also present the different maps calculated between the normalized seismicity rates (linear scale) of the adaptive and fixed corrected models (as Adaptive_corrected − Fixed_corrected), along with the events of the testing catalog, in the same zones of Figure 2: Indonesia (Figure 2a), Mexico (Figure 2b), and Chile (Figure 2c). Colors in light blue to red represent positive differences (i.e., the rate of the adaptive model is higher with respect to the fixed model), deep blue represents negative differences (i.e., the rate of the adaptive model is lower with respect to the fixed model), and blue represents no difference.

Table 5. Log-likelihood differences (ΔLL) between the models.

Models	Magnitude for the Comparison	Log-Likelihood Difference, ΔLL
Corrected Adaptive vs. Adaptive	5.5+	7
Corrected Adaptive vs. Adaptive	6.5+	7
Corrected Fixed vs. Fixed	5.5+	99
Corrected Fixed vs. Fixed	6.5+	18
Adaptive vs. Fixed	5.5+	1658
Adaptive vs. Fixed	6.5+	92

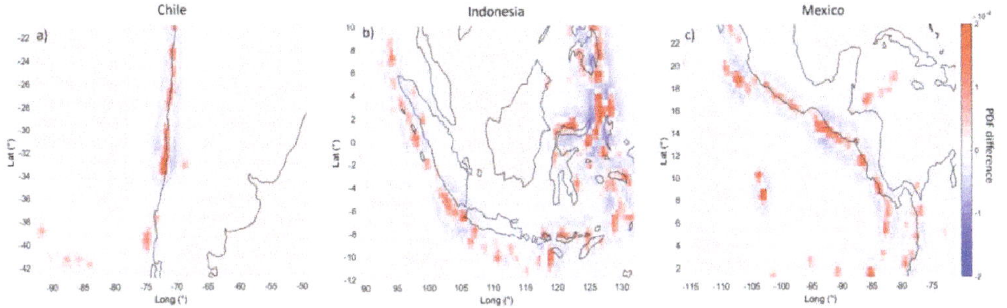

Figure 5. The difference between the normalized seismicity rates (linear scale) of the adaptive and fixed corrected models (Adaptive_corrected − Fixed_corrected) in some zones in the world: Chile (**a**), Indonesia (**b**), and Mexico (**c**).

6. Discussion

The comparison of the four global models, fixed and adaptive smoothed seismicity with and without our correction for the inclusion of aftershocks and foreshocks, clearly shows better performance of the models that use the correction. This positive result indicates that using all events of a seismic sequence instead of only the mainshock increases the forecasting capabilities of the smoothing seismicity models. Another very interesting result is the better performance of the adaptive approach concerning the fixed approach, here demonstrated for a global catalog and two different magnitude thresholds, M_w 5.5+ and 6.5+. Looking at the normalized seismicity rates in Figure 4a–d, it is possible to note the larger smoothing for the fixed models compared to the adaptive models in the zones where the seismicity is higher. The difference between the adaptive and fixed smoothing approaches is evidenced in Figure 5: the large smoothing for the fixed model leads to lower rates with respect to the adaptive model in the zones where the earthquake rate for the testing catalog is higher (pink and red colors in Figure 5). On the contrary, the rates of the fixed model are higher with respect to the adaptive model in the areas adjacent to the more seismic active zones (blue colors in Figure 5). Zones far from the main seismic regions (e.g., intraplate zones with very few earthquakes) have a very small difference between the fixed and adaptive seismicity rate models (light pink color in Figure 5).

The significantly better performances obtained by the adaptive smoothed approach (Table 5) confirm at a global scale the regional results obtained by [4] for California and [31] for Italy. Our method is more straightforward than that of [10], because it does not require a sophisticated stochastic declustering procedure [12]. Still, it only needs to identify the events in a seismic sequence, in this work made with the classical [16] declustering algorithm. Despite its simplicity, our method gives encouraging good results. A possible future work could be a comparison between our approach and the [10] approach.

Our method is based on the assumption of stationarity of the seismicity (usually accepted in long-term modeling); however, working in smaller time and spatial scales, some regions may exhibit different spatiotemporal variations, useful to forecast stronger seismic events [32,33]. Abandoning the stationarity assumption, smaller earthquakes can also be used to try to determine the current state of the seismic cycle [34] and then identify possible temporal variations in the long-term seismic rates.

7. Conclusions

The ten-year, global, pseudoprospective earthquake spatial forecasting experiment gives us two critical results:

(1) In general, the adaptive smoothing approach has better performance with respect to the fixed smoothing approach also for a global catalog with large events ($M_w \geq 5.5$ and $M_w \geq 6.5$);
(2) Using the simple correction described in this work, the inclusion of aftershocks and foreshocks leads to better spatial performances of the smoothed seismicity models.

A possible future improvement of our method is to include the events below the magnitude of completeness ($M_w < 5.5$) in the model to enhance and better describe the active fault structures and their segments.

Author Contributions: M.T. conceived the method; M.T. and A.A. defined the application; M.T. performed the data analysis and created the figures; M.T. and A.A. wrote the paper. Both authors have read and agreed to the published version of the manuscript.

Funding: This study was supported by Centro di Pericolosita' Sismica (CPS), Istituto Nazionale di Geofisica e Vulcanologia (INGV).

Institutional Review Board Statement: Not applicable.

Informed Consent Statement: Not applicable.

Data Availability Statement: Data and code used in this paper are available at: https://github.com/MatteoTaroniINGV/SmoothedSeismicity.

Acknowledgments: This study is under the framework of the Mappa di Pericolosita' Sismica, MPS16 Project supported by Centro di Pericolosita' Sismica (CPS), Instituto Nazionale di Geofisica e Vulcanologia (INGV).

Conflicts of Interest: The authors declare no conflict of interest.

References

1. Electric Power Research Institute. *Seismic Hazard Methodology for the Central and Eastern United States*; EPR/Report NP-4726; Electric Power Research Institute: Palo Alto, CA, USA, 1986; Volume 10.
2. Meletti, C.; Galadini, F.; Valensise, G.; Stucchi, M.; Basili, R.; Barba, S.; Vannucci, G.; Boschi, E. A seismic source zone model for the seismic hazard assessment of the Italian territory. *Tectonophysics* **2008**, *450*, 85–108. [CrossRef]
3. Frankel, A. Mapping seismic hazard in the central and eastern United States. *Seismol. Res. Lett.* **1995**, *66*, 8–21. [CrossRef]
4. Helmstetter, A.; Kagan, Y.Y.; Jackson, D.D. High-resolution time-independent grid based forecast for M $\geq$ 5 earthquakes in California. *Seismol. Res. Lett.* **2007**, *78*, 78–86. [CrossRef]
5. Cornell, C.A. Engineering seismic risk analysis. *Bull. Seismol. Soc. Am.* **1968**, *58*, 1583–1606. [CrossRef]
6. Marzocchi, W.; Taroni, M. Some thoughts on declustering in probabilistic seismic-hazard analysis. *Bull. Seismol. Soc. Am.* **2014**, *104*, 1838–1845. [CrossRef]
7. Le Cam, L. An approximation theorem for the Poisson binomial distribution. *Pac. J. Math.* **1960**, *10*, 1181–1197. [CrossRef]
8. Serfling, R.J. A general Poisson approximation theorem. *Ann. Prob.* **1975**, *3*, 726–731. [CrossRef]
9. Taroni, M.; Akinci, A. Good practices in PSHA: Declustering, b-value estimation, foreshocks and aftershocks inclusion; a case study in Italy. *Geophys. J. Int.* **2021**, *224*, 1174–1187. [CrossRef]
10. Wang, Q.; Jackson, D.D.; Kagan, Y.Y. California earthquake forecasts based on smoothed seismicity: Model choices. *Bull. Seismol. Soc. Am.* **2011**, *101*, 1422–1430. [CrossRef]
11. Ogata, Y. Space-time point-process models for earthquake occurrences. *Ann. Inst. Stat. Math.* **1998**, *50*, 379–402. [CrossRef]
12. Zhuang, J.; Ogata, Y.; Vere-Jones, D. Stochastic declustering of space-time earthquake occurrences. *J. Am. Stat. Assoc.* **2002**, *97*, 369–380. [CrossRef]
13. Ekström, G.; Nettles, M.; Dziewonski, A.M. The global CMT project 2004–2010: Centroid-moment tensors for 13,017 earthquakes. *Phys. Earth Planet. Inter.* **2012**, *200–201*, 1–9. [CrossRef]
14. Dziewonski, A.M.; Chou, T.A.; Woodhouse, J.H. Determination of earthquake source parameters from waveform data for studies of global and regional seismicity. *J. Geophys. Res. Solid Earth* **1981**, *86*, 2825–2852. [CrossRef]
15. Schorlemmer, D.; Wiemer, S.; Wyss, M. Variations in earthquake-size distribution across different stress regimes. *Nature* **2005**, *437*, 539–542. [CrossRef] [PubMed]
16. Gardner, J.K.; Knopoff, L. Is the sequence of earthquakes in Southern California, with aftershocks removed, Poissonian? *Bull. Seismol. Soc. Am.* **1974**, *64*, 1363–1367. [CrossRef]
17. Wiemer, S. A software package to analyze seismicity: ZMAP. *Seismol. Res. Lett.* **2001**, *72*, 373–382. [CrossRef]
18. Hiemer, S.; Woessner, J.; Basili, R.; Danciu, L.; Giardini, D.; Wiemer, S. A smoothed stochastic earthquake rate model considering seismicity and fault moment release for Europe. *Geophys. J. Int.* **2014**, *198*, 1159–1172. [CrossRef]
19. Akinci, A. HAZGRIDX: Earthquake forecasting model for ML C 5.0 earthquakes in Italy based on spatially smoothed seismicity. *Ann. Geophys.* **2010**, *53*, 51–61. [CrossRef]
20. Lombardi, A.M.; Marzocchi, W. The assumption of Poisson seismic-rate variability in CSEP/RELM experiments. *Bull. Seismol. Soc. Am.* **2010**, *100*, 2293–2300. [CrossRef]
21. Kagan, Y.Y. *Earthquakes: Models, Statistics, Testable Forecasts*; John Wiley & Sons: Hoboken, NJ, USA, 2013.
22. Schorlemmer, D.; Gerstenberger, M.C.; Wiemer, S.; Jackson, D.D.; Rhoades, D.A. Earthquake likelihood model testing. *Seismol. Res. Lett.* **2007**, *78*, 17–29. [CrossRef]
23. Schorlemmer, D.; Werner, M.; Marzocchi, W.; Jordan, T.H.; Ogata, Y.; Jackson, D.D.; Mak, S.; Rhoades, D.A.; Gerstenberger, M.C.; Hirata, N.; et al. The collaboratory for the study of earthquake predictability: Achievements and priorities. *Seismol. Res. Lett.* **2018**, *89*, 1305–1313. [CrossRef]
24. Savran, W.H.; Werner, M.J.; Marzocchi, W.; Rhoades, D.A.; Jackson, D.D.; Milner, K.; Field, E.; Michael, A. Pseudoprospective Evaluation of UCERF3-ETAS Forecasts during the 2019 Ridgecrest Sequence. *Bull. Seismol. Soc. Am.* **2020**, *110*, 1799–1817. [CrossRef]
25. Taroni, M.; Zechar, J.D.; Marzocchi, W. Assessing annual global M 6+ seismicity forecasts. *Geophys. J. Int.* **2014**, *196*, 422–431. [CrossRef]
26. Strader, A.; Werner, M.; Bayona, J.; Maechling, P.; Silva, F.; Liukis, M.; Schorlemmer, D. Prospective evaluation of global earthquake forecast models: 2 yrs of observations provide preliminary support for merging smoothed seismicity with geodetic strain rates. *Seismol. Res. Lett.* **2018**, *89*, 1262–1271. [CrossRef]
27. Kass, R.E.; Raftery, A.E. Bayes factors. *J. Am. Stat. Assoc.* **1995**, *90*, 773–795. [CrossRef]
28. Marzocchi, W.; Zechar, J.D.; Jordan, T.H. Bayesian forecast evaluation and ensemble earthquake forecasting. *Bull. Seismol. Soc. Am.* **2012**, *102*, 2574–2584. [CrossRef]
29. Taroni, M.; Zhuang, J.; Marzocchi, W. High-Definition Mapping of the Gutenberg–Richter b-Value and Its Relevance: A Case Study in Italy. *Seismol. Res. Lett.* **2021**, *92*, 3778–3884. [CrossRef]
30. Taroni, M.; Vocalelli, G.; De Polis, A. Gutenberg–Richter B-Value Time Series Forecasting: A Weighted Likelihood Approach. *Forecasting* **2021**, *3*, 561–569. [CrossRef]
31. Akinci, A.; Moschetti, M.P.; Taroni, M. Ensemble smoothed seismicity models for the new Italian probabilistic seismic hazard map. *Seismol. Res. Lett.* **2018**, *89*, 1277–1287. [CrossRef]

32. Sarlis, N.V.; Skordas, E.S.; Varotsos, P.A. Order parameter fluctuations of seismicity in natural time before and after mainshocks. *EPL* **2010**, *91*, 59001. [CrossRef]
33. Sarlis, N.V.; Skordas, E.S.; Varotsos, P.A.; Nagao, T.; Kamogawa, H.; Uyeda, S. Spatiotemporal variations of seismicity before major earthquakes in the Japanese area and their relation with the epicentral locations. *Proc. Natl. Acad. Sci. USA* **2015**, *112*, 986–989. [CrossRef] [PubMed]
34. Rundle, J.B.; Turcotte, D.L.; Donnellan, A.; Grant Ludwig, L.; Luginbuhl, M.; Gong, G. Nowcasting earthquakes. *Earth Space Sci.* **2016**, *3*, 480–486. [CrossRef]

Article

Estimation of the Tapered Gutenberg-Richter Distribution Parameters for Catalogs with Variable Completeness: An Application to the Atlantic Ridge Seismicity

Matteo Taroni [1,*], Jacopo Selva [2] and Jiancang Zhuang [3]

[1] Istituto Nazionale di Geofisica e Vulcanologia (INGV), 00143 Roma, Italy
[2] Istituto Nazionale di Geofisica e Vulcanologia (INGV), 40100 Bologna, Italy; jacopo.selva@ingv.it
[3] The Institute of Statistical Mathematics, Tokyo 190-0014, Japan; zhuangjc@ism.ac.jp
* Correspondence: matteo.taroni@ingv.it

Abstract: The use of the tapered Gutenberg-Richter distribution in earthquake source models is rapidly increasing, allowing overcoming the definition of a hard threshold for the maximum magnitude. Here, we expand the classical maximum likelihood estimation method for estimating the parameters of the tapered Gutenberg-Richter distribution, allowing the use of a variable through-time magnitude of completeness. Adopting a well-established technique based on asymptotic theory, we also estimate the uncertainties relative to the parameters. Differently from other estimation methods for catalogs with a variable completeness, available for example for the classical truncated Gutenberg-Richter distribution, our approach does not need the assumption on the distribution of the number of events (usually the Poisson distribution). We test the methodology checking the consistency of parameter estimations with synthetic catalogs generated with multiple completeness levels. Then, we analyze the Atlantic ridge seismicity, using the global centroid moment tensor catalog, finding that our method allows better constraining distribution parameters, allowing the use more data than estimations based on a single completeness level. This leads to a sharp decrease in the uncertainties associated with the parameter estimation, when compared with existing methods based on a single time-independent magnitude of completeness. This also allows analyzing subsets of events, to deepen data analysis. For example, separating normal and strike-slip events, we found that they have significantly different but well-constrained corner magnitudes. Instead, without distinguishing for focal mechanism and considering all the events in the catalog, we obtain an intermediate value that is relatively less constrained from data, with an open confidence region.

Keywords: statistical methods; statistical seismology; magnitude-frequency distribution; corner magnitude; tapered Pareto; tapered Gutenberg-Richter

Citation: Taroni, M.; Selva, J.; Zhuang, J. Estimation of the Tapered Gutenberg-Richter Distribution Parameters for Catalogs with Variable Completeness: An Application to the Atlantic Ridge Seismicity. *Appl. Sci.* **2021**, *11*, 12166. https://doi.org/10.3390/app112412166

Academic Editors: Stefania Gentili, Rita Di Giovambattista, Robert Shcherbakov, Filippos Vallianatos and Fernando M.S.F. Marques

Received: 9 July 2021
Accepted: 3 December 2021
Published: 20 December 2021

1. Introduction

The Gutenberg-Richter law [1] is the most widely applied magnitude frequency distribution for earthquakes. If we look only to the distribution of the magnitudes, independently from the rate of events, this law corresponds to an exponential distribution [2]. In this case, it depends on only one parameter (the so-called b-value), controlling the slope of the distribution, and does not have an upper bound for the magnitude. In order to have a more physical behavior for the right tail of the magnitude distribution, two other formulations of this law are usually applied: the truncated and the tapered Gutenberg-Richter distributions [3]. The truncated version applies a hard bound to the tail, i.e., a maximum magnitude (M_{max}). Instead, the tapered version applies a soft bound, i.e., a corner magnitude (C_M): the probability of an earthquake bigger than the corner magnitude decreases very rapidly asymptotically reaching zero (see Figure 1).

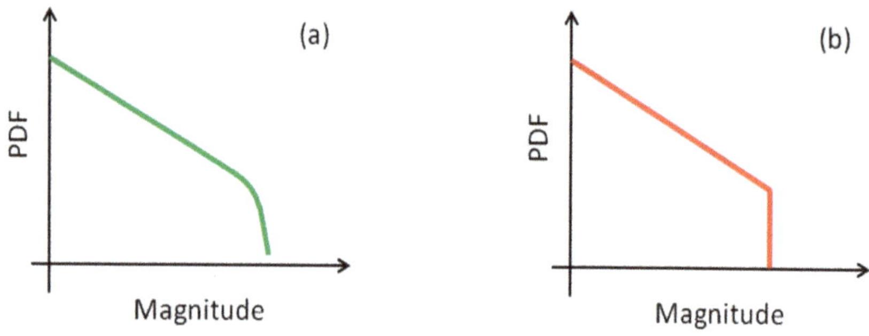

Figure 1. Probability density functions of the tapered (**a**) and truncated (**b**) Gutenberg-Richter distributions, in a log10 Y-axis scale.

For these two formulations of the Gutenberg-Richter distribution, we need an extra parameter to be estimated: the maximum and the corner magnitude for the truncated and the tapered distributions, respectively. Regarding the estimation of the maximum magnitude, Zöller and Holschneider summarized very well the state of the art: "*the earthquake history in a fault zone tells us almost nothing about M_{max}*" [4]. This and other papers [5–7] clearly show that a maximum likelihood estimation (MLE) of M_{max} is not applicable, as the MLE is equal to the maximum observed magnitude and this may be problematic, considering the relatively short observation time as compared with mean recurrence times of large magnitude events. Conversely, the corner magnitude can be properly estimated if a sufficiently large amount of data is available [8]. The tapered Gutenberg-Richter distribution, also called tapered Pareto distribution or "Kagan distribution" by some statistical seismologists, was deeply investigated primarily by Kagan and Schoenberg [8], and then by Kagan [3], Schoenberg and Patel [9], and Geist and Parsons [7]. All these works use seismic catalogs with a single magnitude of completeness. These methods do not need any assumption on the distribution of the number of events. However, the size of the catalog can largely be expanded by adopting multiple levels of completeness, with a completeness magnitude that decreases in time, as the quality and quantity of the available instrumentation improve (Figure 2). This allows including in the estimation both the large number of relatively small events recorded by modern monitoring networks, and the larger events that occurred in the past, possibly also from pre-instrumental times [10].

Existing methods [10–12] that deal with this problem need an assumption regarding the distribution through the time of the events. The distribution usually assumed is the Poisson distribution. This assumption is not always correct for the events in seismic catalogs, in particular if the magnitude of completeness of the catalog is lower than Mw 6.5 [13], forcing the application of declustering algorithms. On the other hand, declustering decreases the number of usable data and may introduce important biases in parameter estimations [14], which may even depend on the declustering algorithm selected.

This paper aims to develop a method to perform the parameters' estimation for catalog with a variable magnitude of completeness (see Figure 2), without making any assumption on the distribution of the number of events. Thus, such a method can take the pros of both the previously described approaches, avoiding the cons relative to the single level of completeness and the Poisson assumption, allowing to use more data in the data estimation.

In the following, we first introduce the method and then we apply it to the Atlantic ridge seismicity. This region is characterized by shallow seismicity with a prevalence of normal/strike-slip mechanisms. The statistics of seismicity for oceanic spreading ridges was already studied in Bird et al. [15] and Bird and Kagan [16], which estimated for these zones a corner magnitude lower than other parts of the world ($C_M \approx 5.8$). Here, focusing on the Atlantic ridge with a longer catalog and our newly developed methodology, we improve the estimation of the parameters of the tapered Gutenberg-Richter distribution exploiting the potentiality of the newly developed method.

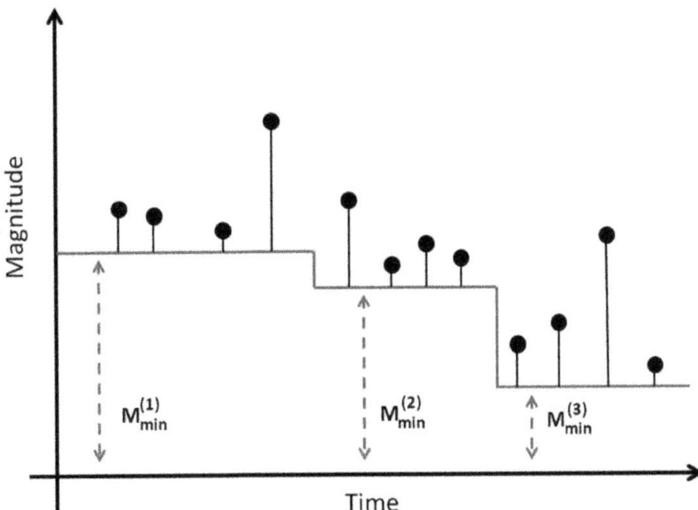

Figure 2. Time vs. magnitude plot for a catalog with a variable magnitude of completeness ($M^{(i)}_{min}$). The grey line represents the completeness, black dots the seismic events.

2. Methods

2.1. Maximum Likelihood Estimation of the Parameters

The Gutenberg-Richter distribution and all its derivations were originally developed using magnitudes. If we use seismic moments (*Mom*) instead of magnitudes, the Gutenberg-Richter distributions (unbounded/truncated/tapered) correspond to the Pareto distributions defined in Kagan [3], with slope parameter β equal to 2/3 the b-value.

The probability density function of the tapered distribution is [3]:

$$f(Mom) = \left(\frac{\beta}{Mom} + \frac{1}{C_{Mom}}\right)\left(\frac{Mom_{min}}{Mom}\right)^{\beta} exp\left(\frac{Mom_{min} - Mom}{C_{Mom}}\right)$$
$$for\ Mom_{min} \leq Mom < \infty \tag{1}$$

where Mom_{min} is the seismic moment of completeness of the catalog, β is the parameter controlling the slope of the distribution, and C_{Mom} is the corner moment that controls the tail of the distribution. We stress that it is always possible to pass from the seismic moment to the magnitude distribution (here, we adopt the relationship defined in Kanamori [17]. In this case the corner moment C_{Mom} is called "corner magnitude" (C_M), and the seismic moment of completeness corresponds to the magnitude of completeness.

If we have a seismic moment of completeness that varies with time ($Mom^{(i)}_{min}$, Figure 2), we can easily rewrite Equation (1) with:

$$f_{(i)}(Mom) = \left(\frac{\beta}{Mom} + \frac{1}{C_{Mom}}\right)\left(\frac{Mom^{(i)}_{min}}{Mom}\right)^{\beta} exp\left(\frac{Mom^{(i)}_{min} - Mom}{C_{Mom}}\right)$$
$$for\ Mom^{(i)}_{min} \leq Mom < \infty \tag{2}$$

This relationship allows referring each observation to the completeness that holds at the time of its occurrence: in this time frame, indeed, Equation (2) describes the statistical distribution that holds.

Being both the parameters of the distribution (C_{Mom} and β) in common to all these distributions ($Mom^{(i)}_{min}$ is a parameter related to the seismic catalog, estimated independently), their likelihood holds for all such distributions. Thus, if we have a seismic catalog with

N earthquakes with moments $x_1, \ldots, x_i, \ldots, x_N$, the log-likelihood of the tapered Pareto distribution becomes:

$$LL(x_1, \ldots, x_N | \beta, C_{Mom}) = \sum_{i=1}^{N} ln\left[f_{(i)}(x_i) \right] \qquad (3)$$

In Equation (3) the probability density function $f_{(i)}$ depends on the seismic moment of completeness relative to the i-th earthquake. In Figure 3 we summarize the scheme of our methodology, applied to completeness thresholds relative to Figure 2: in this case the log-likelihood of Equation (3) is obtained by summing up the log-likelihoods relative to the three different thresholds of completeness. Notably, if the seismic moment of completeness is the same for all the events, Equation (3) becomes the classical log-likelihood for the tapered Pareto distribution [3].

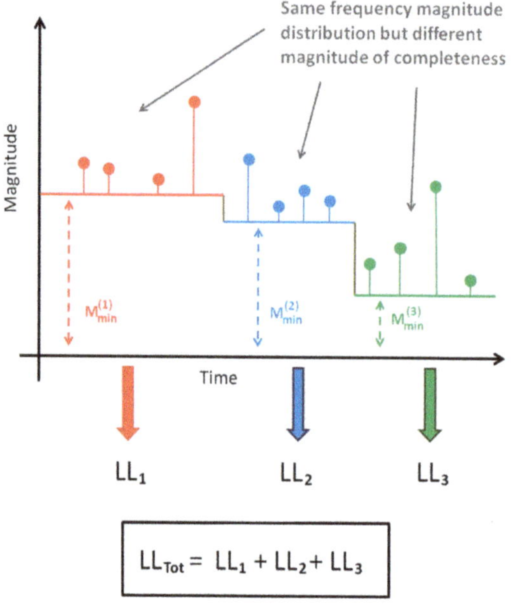

Figure 3. Graphical representation of the log-likelihood computation scheme proposed in this paper in the case of three different magnitudes of completeness thresholds.

To maximize the likelihood of observations, and evaluate the maximum likelihood estimation (MLE) of the parameters β and C_{Mom}, we adopt a brute-force approach, that is, we evaluated the log-likelihood for many potential combinations of the parameters, covering the entire parameter space [12], This allows obtain the complete description of the log-likelihood function: the maximum of the function (LL_{max}) is, by definition, the MLE of the parameters. Moreover, this approach allows evaluating also the shape of the log-likelihood function, which is particularly useful to assess the uncertainty associated with the parameters' estimation, as it will be shown in the next section.

We stress the simplicity of our approach: to move from Equation (1) to Equation (2) we only need to substitute Mom_{min} with $Mom_{min}^{(i)}$, i.e., using the time-variable seismic moment of completeness instead of the fixed one. As the parameters of the distribution that we want to evaluate are in common to all periods, we can simply stack their likelihoods, passing to Equation (3). Noteworthy, this is based on the same principle exploited in Vere-Jones et al. [18], who used a standard log-likelihood function for a tapered Pareto distribution with one or more parameters that change with times (Vere-Jones et al. [18],

Equation (19)), as also in that case, this is possible because each distribution holds at the time of the observation and the different likelihoods can be stacked by summing them (Figure 3).

To check the robustness of this approach, we test its performance by estimating the distribution parameters from synthetic catalogs, for which such parameters are known. To this end, we simulate, using the Taroni and Selva [11] toolbox (based on the Vere-Jones et al. [18] method), thousands of synthetic catalogs with different input parameters (β, C_{Mom}, and magnitude of completeness), obtaining a good agreement between the MLE of the parameters and the input parameters, as expected. The results are shown in Table 1. The goodness of the agreement should be evaluated based on the estimated uncertainty on the parameters. Thus, the results of this comparison are discussed in the next section.

Table 1. Input and estimated β and C_M relative confidence region for thousands of simulated synthetic catalogs.

Number of Simulated Events	Magnitude of Completeness Thresholds	Percentage of Events for Each Completeness	β for the Simulations	Mean of the Estimated β	C_M for the Simulations	Mean of the Estimated C_M	Percentage of Confidence Regions Containing the Values Used in the Simulations
100	5.5; 5.0	50%; 50%	0.67	0.659	6.5	6.467	94.0%
1000	5.5; 5.0	50%; 50%	0.67	0.669	6.5	6.498	95.0%
100	6.0; 5.0	25%; 75%	0.80	0.785	7.5	7.232	93.1%
1000	6.0; 5.0	25%; 75%	0.80	0.798	7.5	7.459	95.2%
100	6.5; 5.3	75%; 25%	0.55	0.546	7.0	6.992	94.9%
1000	6.5; 5.3	75%; 25%	0.55	0.551	7.0	7.001	94.7%

2.2. Estimation of the Uncertainties

To evaluate the uncertainties relative to the parameters' estimation, we use a widely applied method [3,7,8,16,19] based on asymptotic theory [20], sometimes called profile-likelihood confidence region estimation [21]. It states that, if we want to estimate the confidence region of the parameters (in our case, of the tapered Gutenberg-Richter), we have to "cut" the log-likelihood function at a fixed threshold, and then look at the contour plot of this cut. Different thresholds correspond to different confidence intervals. For example, to obtain a 95% confidence region, we have to look at the $LL = LL_{max} - 2.995$ threshold, where LL_{max} is the maximum of the log-likelihood [8]. Hereinafter, the confidence region that describes the uncertainties on the parameters' estimation will be represented by the contour plot of the selected threshold (see Figure 4 for an illustrative example).

This procedure is adopted to evaluate the goodness of the agreement between input and estimated parameters for the thousands of synthetic catalogs discussed in the previous paragraph. In particular, we verify that the input parameters are enclosed in the 95% confidence region for the estimated parameters about 95% of the simulations, obtaining a very good agreement. The results are shown in Table 1.

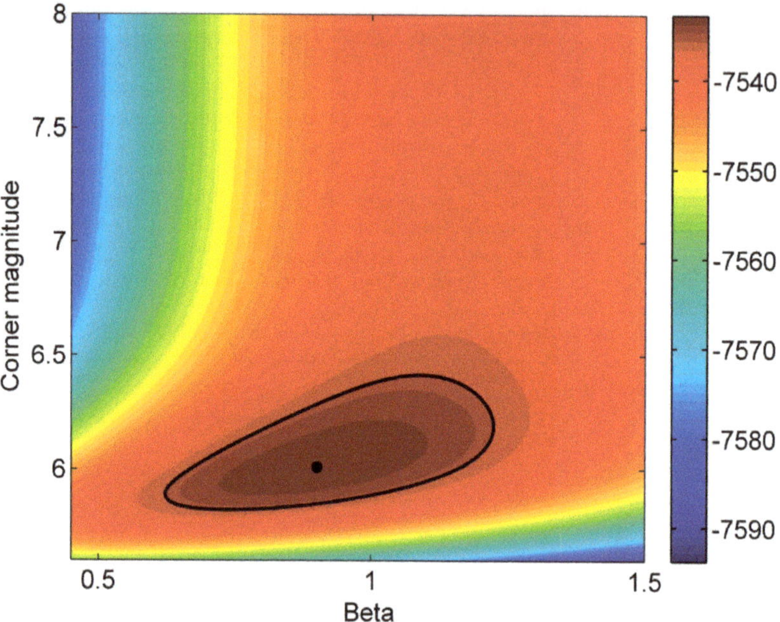

Figure 4. Contour plot of the bivariate log-likelihood function for the parameters of the Tapered Gutenberg-Richter distribution; different colors represent the different log-likelihood values, according to the color bar on the left; the black curve represents the 95% confidence region, and the black dot represents the maximum likelihood estimation MLE.

3. Data

As reference seismic catalog, we use the global centroid moment tensor (CMT) catalog [22,23] of shallow seismicity (depth $\leq$ 50 km) from 1980 to 2019. We do not decluster the catalog, to better exploit the potentiality of our method that does not assume any temporal distribution for earthquakes. As already commented above, this allows using more data and avoiding the introduction of the biases induced by declustering on the β estimation [14,24]. We selected the events in the Atlantic ridge approximately in the latitude range $-60°{:}60°$ (see Figure 5a).

Regarding the magnitude of completeness, we use M_w 5.5 from 1980 and M_w 5.0 from 2004, the ones suggested by the authors of the catalog ([23], see Figure 6). We then carefully test this choice of completeness: as suggested by Marzocchi et al. [25], if the catalog is complete the magnitudes must follow an exponential distribution, and the exponentiality of the magnitudes can be tested through the Lilliefors [26] test. To apply this test with multiple completeness levels, we can build a vector of variables by subtracting to each magnitude the corresponding magnitude of completeness ($M - M_C$), and test the exponentiality of this dataset [27]. The hypothesis of exponential distribution cannot be rejected at any confidence levels, as we obtain a very large p-value (0.50). This demonstrates the robustness of the chosen magnitudes of completeness. A further check is also performed in Figure 5b by plotting the $M - M_C$ vs. the sequential number of events: as suggested by Zhuang et al. [28], a homogeneous pattern near the Y-axis (as it is possible to see in Figure 1b) suggests the correct selection of completeness values for the catalog. The final catalog contains 1168 events.

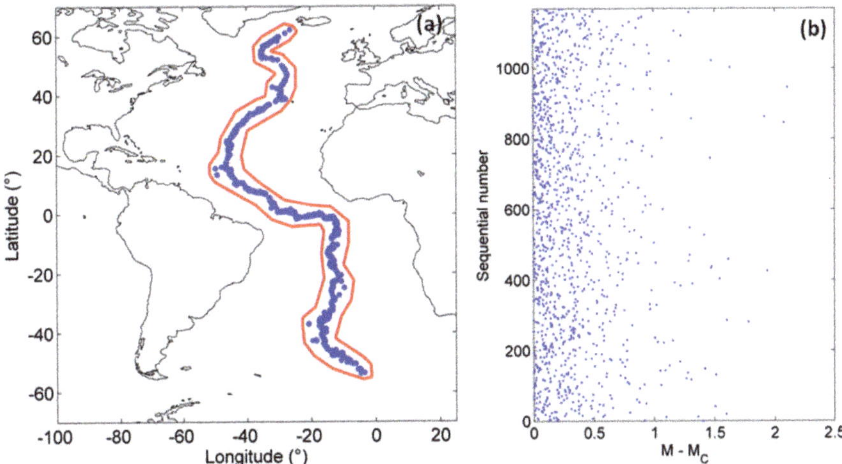

Figure 5. Panel (**a**): events in the Atlantic ridge selected from the CMT catalog (blue dots) inside the red polygon; panel (**b**): difference between the magnitude of events and the relative completeness ($M - M_C$) vs. the sequential number of events.

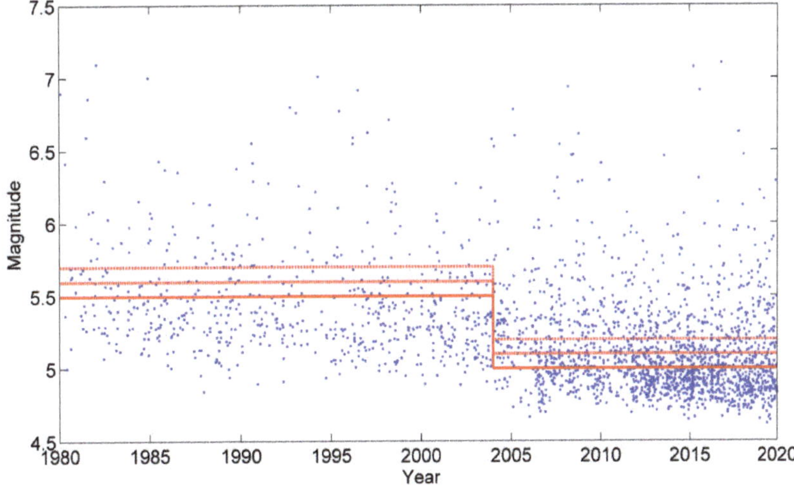

Figure 6. Time vs. magnitude plot; blue dots represent seismic events, red lines the magnitude of completeness thresholds; the solid red line represents the reference level (M_w 5.5 from 1980 and M_w 5.0 from 2004), dashed red lines represent the conservative levels (+0.1 and +0.2 on the reference level).

4. Results

We estimate the corner magnitude and the β of the tapered Pareto distribution both for the whole catalog and for two sub-catalogs: the one containing only normal events and the one containing only the strike-slip events. To select the event in the sub-catalogs, we use the classical Aki-Richards convention for rake: we consider as normal the events with the rake of both nodal planes of the CMT catalog in the range from $-45°$ to $-135°$, and as strike-slip the events with the rake of both nodal planes of the CMT catalog in the range from $-45°$ to $45°$ or $135°$ to $180°$ or $-180°$ to $-135°$. When the two nodal planes have different classifications, the event is not classified. The results of this classification are reported in Table 2. Notably, thrust and undefined events, not contained in either subcatalog, represent only a small part of the events. We underline that in our computation we do not take into account possible uncertainties in the focal mechanism estimation of the CMT catalog; future development of the method will try to introduce these uncertainties in the estimation process.

Table 2. Number of events, percentage (over the whole catalog), and maximum observed magnitude for the different sub-catalogs.

Type	Number of Events	Percentage	Maximum Observed Magnitude
Whole catalog	1168	100%	7.10
Normal events	595	50.9%	6.14
Strike-slip events	523	44.8%	7.10
Thrust events	27	2.3%	6.31
Undefined	23	2.0%	5.83

In Figure 7 we show the results of the estimation for the whole catalog (black curve and dot), for the normal events (green curve and dot), and strike-slip events (red curve and dot); the curves represent the estimated 95% confidence regions (corresponding to 2 standard deviations in normal distributions), while the dots represent the MLE. In the case of distributions with two parameters, the confidence intervals became confidence regions (see Figure 4), to properly capture the 2D nature of these uncertainties.

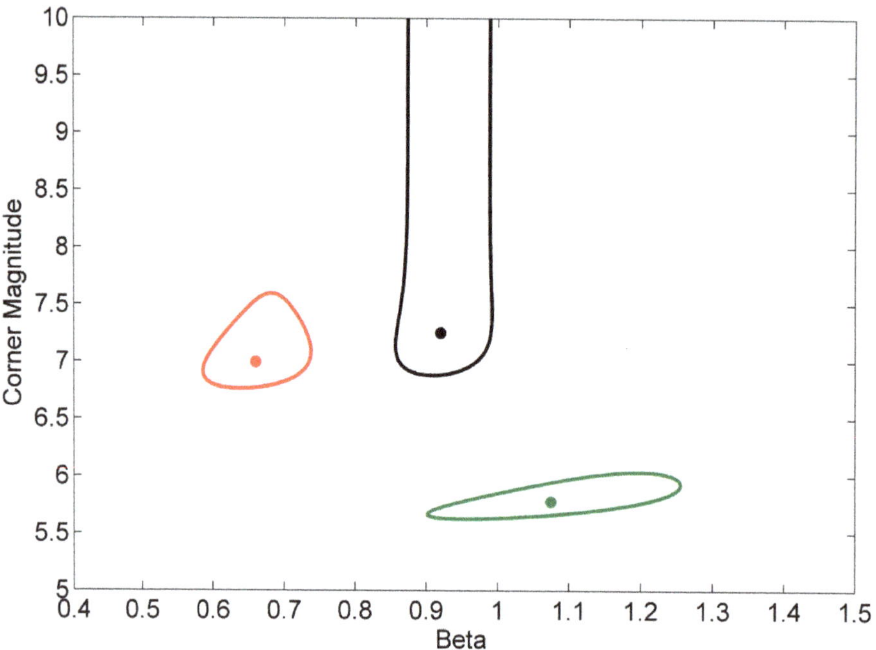

Figure 7. 95% Confidence region estimation (curves) and maximum likelihood estimations MLEs (dots) for the whole catalog (black), and the sub-catalogs with strike-slip events (red), and with normal events (green).

Looking at the shape of the confidence regions, it is evident that the two parameters result fairly uncorrelated. Both for normal and strike-slip events we obtain closed confidence regions, i.e., the confidence regions define a finite area for the uncertainty, showing a well-constrained estimation for all parameters; conversely, for the whole catalog, the confidence region is open toward large corner magnitudes, indicating an unconstrained estimation of the corner magnitude [7,8]. These results are compatible with an infinite corner magnitude corresponding to an unbounded Gutenberg-Richter. We also obtain a clear distinction of the β values for the two sub-catalogs, which results averaged when the whole catalog is used. In Table 3 we show all the MLE for the corner magnitude and β parameters.

Table 3. Maximum likely estimation MLE of the corner magnitude and the slope β, for the whole catalog and for the two sub-catalogs.

Type	Corner Magnitude (MLE)	β (MLE)
Whole catalog	7.25	0.92
Normal events	5.78	1.08
Strike-slip events	7.01	0.66

5. Discussion

By adopting the newly developed procedure, we can consider a much larger dataset for estimating the parameters of the tapered Gutenberg-Richter distribution, as catalog should not be declustered and different magnitude of completeness can be adopted in an older time, extending the temporal coverture of the catalog. This allows a deeper analysis of the Gutenberg-Richter distribution, also considering possible variations in sub-catalogs.

We applied this principle to the seismicity of the Atlantic ridge, obtaining a much-improved description of its seismicity. In particular, the different shapes of the confidence regions obtained considering the whole catalog and the sub-catalogs clearly demonstrate that a mixture of different types of events (i.e., with different focal mechanisms) with different statistical properties for the Gutenberg-Richter distribution can lead to an untrustworthy estimation of its parameters, artificially enlarging their confidence bounds, in particular for the corner magnitude.

We find instead that the corner magnitude for both normal and strike-slip events is well constrained, and incompatible with an unbounded Guttenberg-Richter distribution. This is in agreement with the observation that the size of such a structure is rather limited in the case of oceanic ridges [15]. The estimation of the corner magnitude for the normal event is particularly low (C_M = 5.78), but it is in line with the estimation obtained by Bird et al. [15] for oceanic spreading ridge earthquakes (C_M = 5.83). Conversely, the undifferentiated catalog provides an averaged corner magnitude (biased with respect to both sub-catalogs), with an open confidence region that is compatible with unbounded distribution. Notably, the sub-catalogs almost completely cover the entire catalog, and the thrust and unclassified events not only represent a small subset of events, but also cannot influence the estimation of the corner magnitude, as the maximum observed magnitude for these events is considerably smaller than the one of the whole catalog (6.31 vs. 7.10).

The slope parameter β for strike-slip event is similar to the one of Schorlemmer et al. [29] for the global catalog; on the contrary, the β for the normal events is very high (1.08), corresponding to a b-value equal to 1.62; however, this estimation is quite uncertain (see Figure 7, green curve), with the 95% confidence region for β ranging from 0.90 to 1.25. This large confidence region is compatible with the β estimated by Bird and Kagan [16] for the normal event in the oceanic spreading ridge ($\beta = 0.91$).

The results are pretty independent of the selected completeness. In the Supplementary Material, we perform the same estimation shown in Figure 7, but using a more conservative magnitude of completeness, obtaining very similar results, and thus demonstrating that these results are robust and do not depend on the chosen completeness thresholds.

As discussed above, the newly developed estimation method allows increasing the input dataset by not requiring declustering and by allowing a variation through time of the completeness level. While the largest reduction is due to no declustering, we not that also the possibility of considering different completeness levels has a considerable impact. In Figure 8 we show the specific impact of the use of different completeness thresholds through time, allowed by the newly developed estimation method. The results are compared with the classical estimation method, in which one level of completeness for the whole catalog is used (Mw 5.5 from 1980 to 2019). The lower number of events available using only one level of completeness leads to larger confidence regions. In particular, for normal events, the confidence region computed with the classical method (light green curve in Figure 8) is much bigger than the one computed with the new method (green curve in Figure 8). As expected, a larger amount of available information leads to smaller uncertainties in the estimated parameters, especially in this case. Notably, central values (MLEs) also change, correcting potential biases. For example, the MLE for the entire catalog results outside the confidence bounds defined using more data.

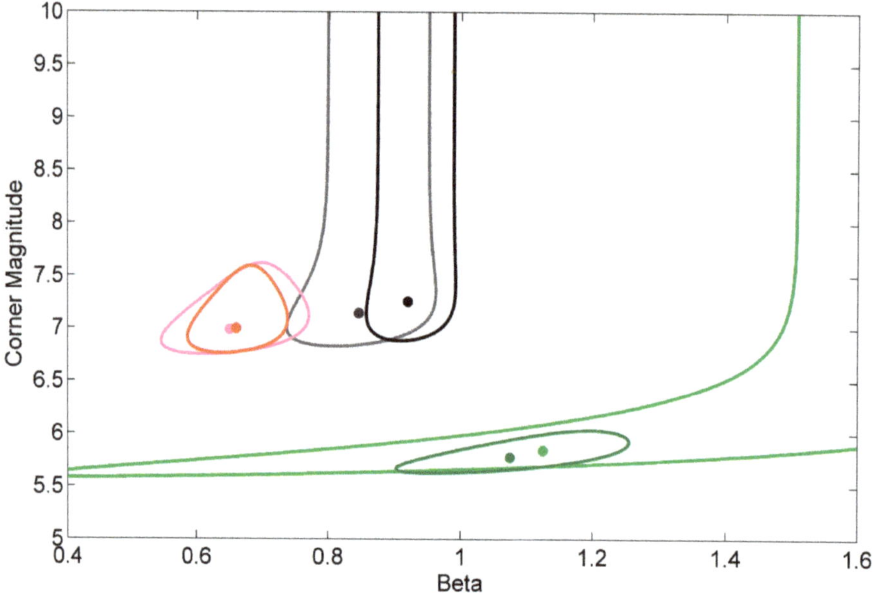

Figure 8. 95% Confidence region estimation (curves) and maximum likelihood estimations MLEs (dots) for the classical estimation approach (light colors: gray, pink, and light green) and the new estimation approach (dark colors: black, red, and green). As in Figure 7, we report both results using for the whole catalog (gray and black), and the sub-catalogs with strike-slip events (pink and red), and with normal events (light green and green).

The different shapes of the confidence region considering the whole catalog and the sub-catalogs show the importance of separating the contribution of different classes of earthquakes to correctly interpret their behavior. Indeed, the averaged behavior estimated from the complete catalog is substantially incompatible with the actual behavior of each single seismicity class. Indeed, by applying the global statistics (obtained by the full catalog) to each individual class, we would implement the wrong statistics to different classes, impacting the hazard in a different way. In our case study for the Atlantic ridge, we would artificially increase the probability of high magnitude normal events. On the contrary, compared with strike-slip events, we demonstrated that normal earthquakes have a significantly smaller corner magnitude coupled with a significantly larger b-value, resulting in a smaller probability of high magnitude normal events.

This not only may complicate the interpretation of the parameter estimation, but also may have a significant impact on hazard quantifications. For example, most of the recent ground motion prediction equations to estimate the attenuation of seismic waves from the source to target are dependent on faulting mechanisms, applying different attenuation laws to the different mechanisms (e.g., [30]). This is probably even more impacting is tsunami hazard, where different mechanisms have a different capability of deforming the sea bottom, resulting in different tsunamigenic capabilities (e.g., [31]). For example, normal events are typically more tsunamigenic than strike-slip events. For not introducing artificial bias in hazard quantification, it will be therefore fundamental to individuate potential mechanism-dependent variation of earthquake statistics and apply hazard models allowing for the aggregation of multiple classes of seismicity (e.g., [32]).

6. Conclusions

The main findings of this work can be summarized by the following two points:

(1) We introduce a new method to estimate the parameters of the tapered Gutenberg-Richter distribution and their uncertainties in the case of catalogs with a variable through-time magnitude of completeness;

(2) We apply this method to the Atlantic ridge seismicity, finding a clear distinct behavior both for the parameters β and corner magnitude, depending on the faulting mechanism: larger β and smaller corner magnitude for normal events, smaller β and larger corner magnitude for strike-slip events.

Supplementary Materials: The following are available online at https://www.mdpi.com/article/10.3390/app112412166/s1. Figure S1: 95% confidence region estimation (curves) and MLE (dots) for the whole catalog (black), strike-slip events (red), and normal events (green) (completeness thresholds +0.1); Figure S2: 95% confidence region estimation (curves) and MLE (dots) for the whole catalog (black), strike-slip events (red), and normal events (green) (completeness thresholds +0.2).

Author Contributions: M.T. and J.Z. conceived the method; M.T. and J.S. defined the application; M.T. performed the data analysis and created the figures; M.T. and J.S. wrote the paper. All authors have read and agreed to the published version of the manuscript.

Funding: This work benefited of the agreement between Istituto Nazionale di Geofisica e Vulcanologia and the Italian Presidenza del Consiglio dei Ministri, Dipartimento della Protezione Civile (DPC), and of the project "Assessment of Cascading Events triggered by the Interaction of Natural Hazards and Technological Scenarios involving the release of Hazardous Substances", funded by the Italian Ministry MIUR PRIN (Progetti di Ricerca di Rilevante Interesse Nazionale) 2017—Grant 2017CEYPS8.

Institutional Review Board Statement: Not applicable.

Informed Consent Statement: Not applicable.

Data Availability Statement: Data and code used in this paper are available at: https://github.com/MatteoTaroniINGV/TaperedGR_Estimation_Catalogs_with_variable_completeness (accessed on 8 July 2021).

Acknowledgments: The authors thank the two anonymous reviewers for their comments which greatly improved the paper.

Conflicts of Interest: Authors declare no conflict of interest.

Abbreviations

Abbreviation	Meaning
M_{max}	Maximum magnitude
MLE	Maximum likelihood estimation
M_{min}	Magnitude of completeness
CMT	Centroid moment tensor catalog
C_M	Corner magnitude

References

1. Gutenberg, B.; Richter, C.F. Frequency of earthquakes in California. *Bull. Seismol. Soc. Am.* **1944**, *34*, 185–188. [CrossRef]
2. Aki, K. Maximum likelihood estimate of b in the formula logN = a − bM and its confidence limits. *Bull. Earthq. Res. Inst.* **1965**, *43*, 237–239.
3. Kagan, Y.Y. Seismic moment distribution revisited: I. Statistical results. *Geophys. J. Int.* **2002**, *148*, 520–541. [CrossRef]
4. Zöller, G.; Holschneider, M. The earthquake history in a fault zone tells us almost nothing about mmax. *Seismol. Res. Lett.* **2016**, *87*, 132–137. [CrossRef]
5. Holschneider, M.; Zöller, G.; Hainzl, S. Estimation of the maximum possible magnitude in the framework of a doubly truncated Gutenberg–Richter model. *Bull. Seismol. Soc. Am.* **2011**, *101*, 1649–1659. [CrossRef]
6. Zöller, G.; Holschneider, M.; Hainzl, S. The maximum earthquake magnitude in a time horizon: Theory and case studies. *Bull. Seismol. Soc. Am.* **2013**, *103*, 860–875. [CrossRef]
7. Geist, E.L.; Parsons, T. Undersampling power-law size distributions: Effect on the assessment of extreme natural hazards. *Nat. Hazards* **2014**, *72*, 565–595. [CrossRef]
8. Kagan, Y.Y.; Schoenberg, F. Estimation of the upper cutoff parameter for the tapered Pareto distribution. *J. Appl. Probab.* **2001**, *38*, 168–185. [CrossRef]
9. Schoenberg, F.P.; Patel, R.D. Comparison of Pareto and tapered Pareto distributions for environmental phenomena. *Eur. Phys. J. Spec. Top.* **2012**, *205*, 159–166. [CrossRef]
10. Kijko, A.; Sellevoll, M.A. Estimation of earthquake hazard parameters from incomplete data files. Part I. Utilization of extreme and complete catalogs with different threshold magnitudes. *Bull. Seismol. Soc. Am.* **1989**, *79*, 645–654. [CrossRef]
11. Weichert, D.H. Estimation of the earthquake recurrence parameters for unequal observation periods for different magnitudes. *Bull. Seismol. Soc. Am.* **1980**, *70*, 1337–1346. [CrossRef]
12. Taroni, M.; Selva, J. GR_EST: An OCTAVE/MATLAB Toolbox to Estimate Gutenberg–Richter Law Parameters and Their Uncertainties. *Seismol. Res. Lett.* **2021**, *92*, 508–516. [CrossRef]
13. Kagan, Y.Y. Earthquake number forecasts testing. *Geophys. J. Int.* **2017**, *211*, 335–345. [CrossRef]
14. Taroni, M.; Akinci, A. Good practices in PSHA: Declustering, b-value estimation, foreshocks and aftershocks inclusion; a case study in Italy. *Geophys. J. Int.* **2021**, *224*, 1174–1187. [CrossRef]
15. Bird, P.; Kagan, Y.Y.; Jackson D., D. Plate tectonics and earthquake potential of spreading ridges and oceanic transform faults. In *Plate Boundary Zones*; Stein, S., Freymueller, J.T., Eds.; AGU: Washington, DC, USA, 2002; Volume 30, pp. 203–218.
16. Bird, P.; Kagan, Y.Y. Plate-tectonic analysis of shallow seismicity: Apparent boundary width, beta, corner magnitude, coupled lithosphere thickness, and coupling in seven tectonic settings. *Bull. Seismol. Soc. Am.* **2004**, *94*, 2380–2399. [CrossRef]
17. Kanamori, H. The energy release in great earthquakes. *J. Geophys. Res.* **1977**, *82*, 2981–2987. [CrossRef]
18. Vere-Jones, D.; Robinson, R.; Yang, W. Remarks on the accelerated moment release model: Problems of model formulation, simulation and estimation. *Geophys. J. Int.* **2001**, *144*, 517–531. [CrossRef]
19. Kagan, Y.Y.; Bird, P.; Jackson, D.D. Earthquake patterns in diverse tectonic zones of the globe. *Pure Appl. Geophys.* **2010**, *167*, 721–741. [CrossRef]
20. Wilks, S.S. *Mathematical Statistics*; Wiley: New York, NY, USA, 1962.
21. Venzon, D.J.; Moolgavkar, S.H. A method for computing profile-likelihood-based confidence intervals. *J. R. Stat. Soc. Ser. C Appl. Stat.* **1988**, *37*, 87–94. [CrossRef]
22. Dziewonski, A.M.; Chou, T.-A.; Woodhouse, J.H. Determination of earthquake source parameters from waveform data for studies of global and regional seismicity. *J. Geophys. Res.* **1981**, *86*, 2825–2852. [CrossRef]
23. Ekström, G.; Nettles, M.; Dziewonski, A.M. The global CMT project 2004–2010: Centroid-moment tensors for 13,017 earthquakes. *Phys. Earth Planet. Inter.* **2012**, *200–201*, 1–9. [CrossRef]
24. Mizrahi, L.; Nandan, S.; Wiemer, S. The effect of declustering on the size distribution of mainshocks. *Seismol. Res. Lett.* **2021**, *92*, 2333–2342. [CrossRef]
25. Marzocchi, W.; Spassiani, I.; Stallone, A.; Taroni, M. How to be fooled searching for significant variations of the b-value. *Geophys. J. Int.* **2020**, *220*, 1845–1856. [CrossRef]
26. Lilliefors, H.W. On the Kolmogorov-Smirnov test for the exponential distribution with mean unknown. *J. Am. Stat. Assoc.* **1969**, *64*, 387–389. [CrossRef]
27. Taroni, M. Back to the future: Old methods for new estimation and test of the Gutenberg-Richter b-value for catalogs with variable completeness. *Geophys. J. Int.* **2021**, *224*, 337–339. [CrossRef]
28. Zhuang, J.; Ogata, Y.; Wang, T. Data completeness of the Kumamoto earthquake sequence in the JMA catalog and its influence on the estimation of the ETAS parameters. *Earth Planets Space* **2017**, *69*, 1–12. [CrossRef]
29. Schorlemmer, D.; Wiemer, S.; Wyss, M. Variations in earthquake-size distribution across different stress regimes. *Nature* **2005**, *437*, 539–542. [CrossRef] [PubMed]
30. Douglas, J.; Edwards, B. Recent and future developments in earthquake ground motion estimation. *Earth-Sci. Rev.* **2016**, *160*, 203–219. [CrossRef]

31. Grezio, A.; Babeyko, A.; Baptista, M.A.; Behrens, J.; Costa, A.; Davies, G.; Geist, E.; Glimsdal, S.; Gonzales, F.I.; Griffin, J.; et al. Probabilistic Tsunami Hazard Analysis: Multiple Sources and Global Applications. *Rev. Geophys.* **2017**, *55*. [CrossRef]
32. Selva, J.; Tonini, R.; Molinari, I.; Tiberti, M.M.; Romano, F.; Grezio, A.; Melini, D.; Piatanesi, A.; Basili, R.; Lorito, S. Quantification of source uncertainties in Seismic Probabilistic Tsunami Hazard Analysis (SPTHA). *Geophys. J. Int.* **2016**, *205*, 1780–1803. [CrossRef]

Article

Physics-Based Simulation of Sequences with Foreshocks, Aftershocks and Multiple Main Shocks in Italy

Rodolfo Console [1,2,†], **Paola Vannoli** [1,*,†,] and **Roberto Carluccio** [1,†]

1   Istituto Nazionale di Geofisica e Vulcanologia, 00143 Rome, Italy; rodolfo.console@ingv.it (R.C.);
     roberto.carluccio@ingv.it (R.C.)
2   Center of Integrated Geomorphology for the Mediterranean Area, 85100 Potenza, Italy
*   Correspondence: paola.vannoli@ingv.it
†   These authors contributed equally to this work.

Abstract: We applied a new version of physics-based earthquake simulator upon a seismogenic model of the Italian seismicity derived from the latest version of the Database of Individual Seismogenic Sources (DISS). We elaborated appropriately for their use within the simulator all fault systems identified in the study area. We obtained synthetic catalogs spanning hundreds of thousands of years. The resulting synthetic seismic catalogs exhibit typical magnitude, space and time features that are comparable to those obtained by real observations. A typical aspect of the observed seismicity is the occurrence of earthquake sequences characterized by multiple main shocks of similar magnitude. Special attention was devoted to verifying whether the simulated catalogs include this notable aspect, by the use of an especially developed computer code. We found that the phenomenon of Coulomb stress transfer from causative to receiving source patches during an earthquake rupture has a critical role in the behavior of seismicity patterns in the simulated catalogs. We applied the simulator to the seismicity of the northern and central Apennines and compared the resulting synthetic catalog with the observed seismicity for the period 1650–2020. The result of this comparison supports the hypothesis that the occurrence of sequences containing multiple mainshocks is not just a casual circumstance.

Keywords: numerical modeling; earthquake simulator; statistical methods; earthquake clustering; northern and central Apennines

Citation: Console, R.; Vannoli, P.; Carluccio, R. Physics-Based Simulation of Sequences with Foreshocks, Aftershocks and Multiple Main Shocks in Italy. *Appl. Sci.* **2022**, *12*, 2062. https://doi.org/10.3390/app12042062

Academic Editor: Stefania Gentili

Received: 22 December 2021
Accepted: 8 February 2022
Published: 16 February 2022

Publisher's Note: MDPI stays neutral with regard to jurisdictional claims in published maps and institutional affiliations.

1. Introduction

A typical aspect of the observed seismicity in the northern and central Apennines, and in the whole Italian region more generally, is the occurrence of earthquake sequences characterized by multiple, similarly large mainshocks. An example of this behavior is the quantitative model "Every Earthquake Precursory According to Scale" (EEPAS), applied by Rhoades and Evison [1,2,3]. According to their quantitative definition, introduced by Evison and Rhoades [4], swarms are seismic sequences constituted by at least three earthquakes whose magnitudes are linked to each other by empirical rules.

In this study we define as a multiplet a set of two or more earthquakes, with the following conditions: (a) the first event has a magnitude equal to or larger than a given threshold; (b) the others occur within a time difference and distance defined by the Gardner and Knopoff [5] criterion from each other; and (c) within a given magnitude range. This definition is different from that usually applied for common seismicity patterns such as foreshock–aftershock sequences and clusters (e.g., Gentili and Di Giovambattista [6]).

Building upon a previous paper (Console et al. [7]), in which we examined the aspect of multiple mainshocks in central Italy, in this study we aim at verifying if a synthetic catalog reproduces this kind of earthquake clustering. For this purpose, we apply a new version of the simulator algorithm, in which the role of stress transfer among elements of an expanding rupture is enhanced. Moreover, we give also examples of other space-time seismic features exhibited by synthetic catalogs, both in short- (days–months) and long-term (years–centuries), some of which were observed in real earthquake catalogs.

In Section 2 we present a brief description of the algorithm used for detecting multiple events in an earthquake catalog, based on the previously cited (a), (b), and (c) criteria. This algorithm is applied for providing a possible metric for comparing real observations with simulations.

Section 3 gives an outline of the seismotectonic model of our study area of northern and central Apennines, along with examples of recent and historical sequences of multiple mainshocks observed in this region.

In Section 4, after a short introduction of the new version of the simulator employed in this study, we show the results obtained applying this simulation code to the above mentioned seismotectonic model of the study area. Having tried three choices for the two main free parameters present in the algorithm, for a total of nine different combinations, we chose one of them by a criterion based on the analysis of the multiplets in the synthetic catalog of 100,000 years. Some features of this preferred simulated catalog are then compared in several ways with a real set of observations lasting only 370 years in the same seismogenic area.

Section 5 reports other results of spatio-temporal analysis of the same 100,000 years simulated catalog that appear to be consistent in reasonable way with real seismicity patterns not strictly related to our study area. In particular, we show that the use of simulators allows testing hypotheses of seismogenic models in a way that is not possible on the basis of real observations, due to lack of completeness and homogeneity of these observations in the long-term.

2. The Algorithm for Identification of Multiple Events

A special algorithm for the search of multiple events in a seismic catalog was created to use it as metrics in the comparison between the simulator results and the observed seismicity of the studied region. The computer code is "customer-built" and it was already introduced by Console et al. [7]. There is no specific definition of "sequence with multiple main shocks", nor any fixed magnitude values. We give here a brief description for a better understanding of its use. At its first level, the algorithm systematically analyzes time-ordered couples of events to check if they meet some constitutive conditions. Once matching couples are found, they are then used as elements for ordered noncyclic graph construction. These graphs can be 'traversed' to find in them the searched multiple events groups. In accordance with the above definitions, we developed a method based on four criteria for our comparisons among couples of events (Table 1):

1. There is a minimum magnitude threshold for the first event of the group (hereafter called "pivot");
2. The magnitude of any other main shocks of the sequence must lay in a predefined neighborhood of the pivot's magnitude;
3. The events' time differences must be less than a threshold time, which is a function of event magnitudes (subject to criteria #1 and #2);
4. For any event, a magnitude dependent distance (a radius) is defined and the distance between the epicentres must be smaller than a proper function of those radii.

The values associated to the criteria #1 and #2 are selected by the user and based on needs, expert judgment and/or knowledge of the instrumental–historical seismicity of the area. The relations for time (#3) and distance (#4) thresholds as a function of event magnitudes were derived from Gardner and Knopoff [5] empiric tables and as epicentre distance threshold. Notice that the criterion #3 does not provide for a choice, while the criterion #4 provides for the choice among three different ways of applying the formulas of Gardner and Knopoff [5]. In this study, we used the sum of the radii. Once launched, the algorithm parses the catalog through a cyclic, the three-step analysis procedure is repeated until the end of the file:

- The first step starts with the selection of the next pivot event and the definition of a pool of eligible events (if they exist). They are found using criterion #3;
- The second phase is a thorough analysis of all useful couples taken from the pool, checked for fulfillment of criteria #1, #2 and #4;
- The last step is the construction of the graph, its traversal for the multiplets group search, its eventual output in the output buffer, and the flagging of used events to not reuse them after the next pivot search.

Table 1 summarizes the formulas, explaining the rationale by which they were used in this study, and lists our choices for the threshold values.

Table 1. Constitutive criteria for our clustering analysis algorithm. For each criterion, the second column shows rationale on which it is based, the third column contains formulas, and the last column contains choices used in this paper.

	Criterion	Derived by	Formula	Our Choice
1	Threshold magnitude for the first event (named pivot) (M_{thr})	Expert Judgment	$M_{pivot} \geq M_{thr}$	$M_{thr} = 5.5$
2	Magnitude difference with the pivot ($M_{pivot} - M_{E2}$)	Expert Judgment	$M_{E2} \geq M_{pivot} - a$ $M_{E2} \leq M_{pivot} + b$	$a = 0.5$ $b = 0.5$
3	Time difference between the occurrence of main shocks ($t_{E2} - t_{E1}$)	Empirical Relationship (Gardner and Knopoff [5])	$(t_{E2} - t_{E1}) \leq t_{GK}(M_{E1})$	n/a
4	Spatial distance between hypocenters ($|\vec{x}_{E2} - \vec{x}_{E1}|$)	Empirical Relationship (Gardner and Knopoff [5]) + Expert Judgment	$|\vec{x}_{E2} - \vec{x}_{E1}| \leq$ chosen from: (a) $r_{GK}(M_{E1})$ (b) $\mathrm{Max}[r_{GK}(M_{E1}), r_{GK}(M_{E2})]$ (c) $r_{GK}(M_{E1}) + r_{GK}(M_{E2})$	(c)

Even if the algorithm cannot be called "optimal" in principle, since it is based on an arbitrary choice among possible criteria, it is, however, quite effective, and its importance lays in the metrics it represents for comparison among seismic catalogs.

3. Seismotectonic Model

The seismogenic model of the study area straddles northern and central Italy from the large flat area of the Po Plain (to the north) toward the northern flank of the Gran Sasso mountain range, the highest sector of the Apennines (to the south, Figure 1). The study area is wider than that previously studied (Console et al. [8]) through an old version of the simulator and the seismogenic sources now come from the latest 3.3.0 version of the DISS (DISS Working Group [9]).

Historical and instrumental seismicity in the study area is mainly distributed along the axis of the northern and central Apennines chain and, secondarily, in correspondence with its foothills, plains, and coastal areas (Rovida et al. [10]). The causative sources of the earthquakes of these two regions have different parameters and kinematics, as shown by focal mechanisms (Pondrelli et al. [11]), active stress indicators (Mariucci and Montone [12]), geological data (see DISS Working Group [9] and references therein), and active strain data (Devoti et al. [13]). As matter of fact, the GPS data show the crustal extension at a rate of about 3 mm/yr across the Apennines belt and the compression towards the Adriatic foreland (Devoti et al. [13]).

The active extension along the backbone of the Apennines is accommodated by normal faulting, which dominates along the hinge of the chain at shallow crustal seismogenic depth (blue polygons in Figure 1; e.g., Vannoli et al. [14]). The strongest extensional recent earthquakes in the study area occurred during the 2016-2017 central Italy seismic sequence that struck central Apennines with multiple mainshocks (Table 2). The sequence initiated on 24 August 2016 with the M_w 6.2 Amatrice earthquake and was followed on 26 October 2016 by the M_w 6.1 Visso earthquake, about 25 km to the north. The largest event, the M_w 6.6 Norcia earthquake, occurred on 30 October 2016 and nucleated between the source regions of the two previous mainshocks (e.g., Michele et al. [15]; Rovida et al. [10]). Low-magnitude earthquakes of this sequence still occur today (http://terremoti.ingv.it/ accessed on 22 November 2021). This seismic sequence activated a circa 80 km long, NNW-SSE trending, low-angle multiple fault systems (IDs 127 and 128 in Figure 1). These fault systems exhibit complex ruptures and are the easternmost normal faults of the central Apennines, just west of where compressional activity prevails (e.g., Basili et al. [16]; Bonini et al. [17]; Di Bucci et al. [18]; DISS Working Group [9]).

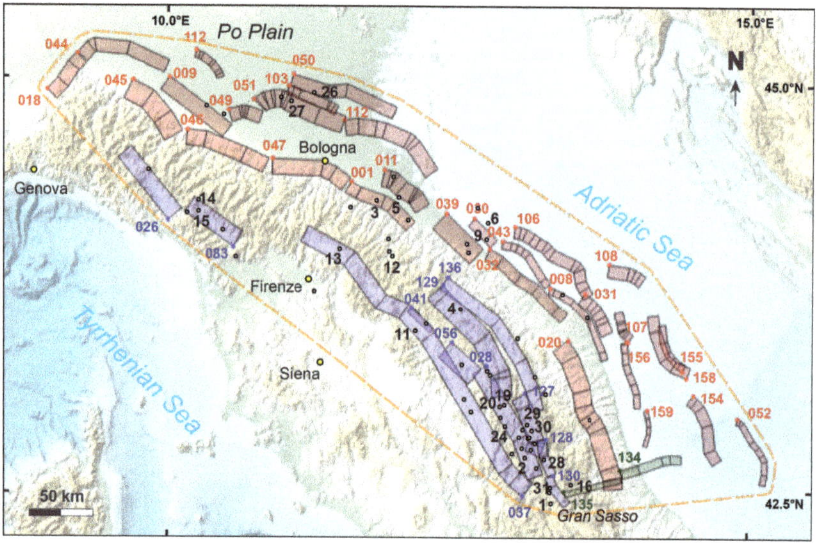

Figure 1. Fault systems and earthquakes. Forty-three DISS (version 3.3.0) seismogenic fault systems are divided into 198 quadrilaterals that best approximate the DISS composite sources, and they are labeled with last three numbers of their DISS-IDs (DISS Working Group [9]). They are shown in accordance with their kinematics (extensional in blue, compressional in red, strike-slip in green), and have colored circles associated with their upper-left corners. Epicentres of the earthquakes from 1650 to 2020 A.D., with $M_w \geq 5.5$, within a 5 km buffer from faults (dotted line) are shown by black circles. Main shocks of Table 2 are labeled in black (Rovida et al. [10]).

The active compression in the Adriatic foreland is mainly accommodated by thrust faulting (e.g., Vannoli et al. [19], Vannoli et al. [20]). Thrust faulting is widespread along the external fronts (red polygons in Figure 1) and propagates from the inner and coastal areas towards the offshore (to the east) and the Po Plain (to the north). Strongest recent compressional earthquakes of the study area occurred in Emilia during the 2012 sequence. This sequence began with the 20 May M_w 6.1 earthquake and was followed on 29 May 2012 by the M_w 5.9 earthquake; therefore, it is characterized by two similarly large mainshocks (see also Figure 1 in Console et al. [7]). The causative faults systems of the 2012 sequence are the external arcs of the most advanced and buried portions of the northern Apennines (IDs 103 and 51 in Figure 1; e.g., Vannoli et al. [20]).

Therefore, the seismogenic model of northern and central Apennines includes onshore and offshore seismogenic sources characterized by both extensional and compressive kinematics (DISS Working Group [9]). In addition, dextral strike-slip faulting is present in the southernmost study area, at the northern border of the Gran Sasso ridge (green polygons in Figure 1). Generally, the transverse structures are faults inherited from older tectonic phases that cut the Adriatic foreland areas accommodating the segmentation of the thrust fronts and the outward propagation of the fold and thrust belts (e.g., Zampieri et al. [21]). Specifically, these strike-slip sources are high-angle, ENE–WSW-trending faults bounding the central Apennines thrust fronts and the southern part of the Apennines basal decollement. They are relatively deep (having 15–20 km of maximum depth), with shear zones that affect the Adriatic foreland (IDs 135 and 134). The western source (ID 135) is believed to be responsible for the seismic sequence that includes two relatively similar large mainshocks that occurred on 5 September 1950 (M_w 5.7) and 8 August 1951 (M_w 5.3; see Table 2).

In summary, the earthquake sequences characterized by at least two similarly large mainshocks are rather common in the study area, affect compressional, extensional, and strike-slip environments, and are very different from the sequences made up of a single large earthquake followed by aftershocks of decreasing magnitude. Figure 2 shows the epicentres of the CPTI15 catalog from 1650 to 2020, with $M_w \geq 5.0$, and the colored lines connect the multiple main shocks events recognized by the algorithm described in the text and reported in column "Csum" of the Table 2. In the same Table the column "C1st" shows the results of the algorithm applying the criterion 4a.

Table 2. Largest sequences in real catalog with at least two main shocks of past 370 years (1650–2020; magnitude and locality from CPTI15). Results of algorithm for detecting sequences with multiple main shocks in study area are shown in last two columns. Column "Csum" shows results of algorithm applying criterion 4c, while column "C1st" criterion 4a (Y: simulated; N: not simulated). Kin: Kinematics; N: normal; S: strike-slip; T: thrust; n.a.: not applicable; * inferred (faults and kinematics responsible for historical earthquakes are inferred)

#	Date	Locality	M_w	Kin	Causative Fault	C1st	Csum
1	14 Jan 1703	Valnerina	6.92	N *	Two main neighboring systems of extensional faults	Y	Y
2	2 Feb 1703	Aquilano	6.67	N *	separated by the Olevano–Antrodoco–Sibillini regional tectonic structure	Y	Y
3	4 Apr 1781	Faentino	6.12	T *	Two distinct fault systems with different current	N	Y
4	3 Jun 1781	Cagliese	6.51	N *	kinematics (two segments of the Pedeapenninic	N	Y
5	17 Jul 1781	Faentino	5.61	T *	thrust front and a segment of the easternmost normal fault system of the northern Apennines)	N	N
6	17 May 1916	Riminese	5.82	T *	The faults responsible for the 1916 sequence are	Y	Y
7	15 Aug 1916	Riminese	5.34	T *	compressive faults close together and located along	Y	Y
8	15 Aug 1916	Riminese	5.35	T *	the coast or immediately offshore. The fault	Y	Y
9	16 Aug 1916	Riminese	5.82	T *	responsible for the 1917 earthquake is an extensional	Y	Y
10	16 Aug 1916	Riminese	5.46	T *	fault located along the backbone of the northern	Y	Y
11	26 Apr 1917	Alta Valtiberina	5.99	N *	Apennines	N	Y
12	10 Nov 1918	Appennino for-livese	5.96	n/a	Distinct extensional fault systems along the backbone of the northern Apennines	Y	Y
13	29 Jun 1919	Mugello	6.38	N *		Y	Y
14	6 Sep 1920	Garfagnana	5.61	N *		N	Y
15	7 Sep 1920	Garfagnana	6.53	N *		N	N
16	5 Sep 1950	Gran Sasso	5.69	S *	The first two events, close to each other, most likely	Y	Y
17	8 Aug 1951	Gran Sasso	5.25	S *	belong to the same transcurrent system (see text). The	Y	Y
18	1 Sep 1951	Monti Sibillini	5.25	n/a	fault responsible for the third event is not known, and it could be a relatively deep source	N	Y
19	26 Sep 1997	Appennino umbro-marchigiano	5.66	N	SW-dipping low-angle normal fault system straddles the central Apennines. The three largest events of the sequence ruptured three adjacent normal fault	Y	Y
20	26 Sep 1997	Appennino umbro-marchigiano	5.97	N	segments	Y	Y

Table 2. *Cont.*

#	Date	Locality	M_w	Kin	Causative Fault	C1st	Csum
21	3 Oct 1997	Appennino umbro-marchigiano	5.22	N		Y	Y
22	6 Oct 1997	Appennino umbro-marchigiano	5.47	N		Y	Y
23	12 Oct 1997	Valnerina	5.19	N		Y	Y
24	14 Oct 1997	Valnerina	5.62	N		Y	Y
25	26 Mar 1998	Appennino umbro-marchigiano	5.26	N		Y	Y
26	20 May 2012	Pianura emiliana	6.09	T	Two parallel fault systems along the most advanced and buried thrusts of the northern Apennines (see text)	Y	Y
27	29 May 2012	Pianura emiliana	5.90	T		Y	Y
28	24 Aug 2016	Amatrice	6.18	N	Multiple fault systems exhibiting complex ruptures along the backbone of the central Apennines (see text)	Y	Y
29	26 Oct 2016	Visso	6.07	N		Y	Y
30	30 Oct 2016	Norcia	6.61	N		Y	Y
31	18 Jan 2017	Aquilano	5.70	N		Y	Y

The seismogenic model upon which we applied the simulator code was derived from the Composite Seismogenic Sources (CSS) of DISS, version 3.3.0 (DISS Working Group [9]). The CSSs are parameterized crustal faults based on regional surface and subsurface geological data, and they are believed to be capable of producing $M_w \geq 5.5$ earthquakes. We converted the 43 CSSs identified in the study area into 198 quadrilaterals specifically developed for this study, and this is consistent with all the geometrical and kinematics parameters supplied for the CSSs (Figure 1). The Table S1 in the Supplementary Material reports the list and the parameters of the 198 quadrilaterals recognized in the study area. Figure S4 shows a sketch of a quadrilateral fault segment and the description of its geometrical parameters.

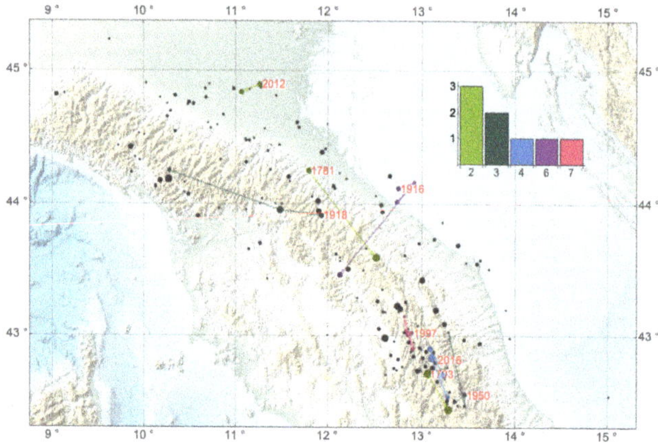

Figure 2. Epicenters of CPTI15 catalog from 1650 to 2020, with $M_w \geq 5.0$. Colored lines join mainshocks of the same sequence recognized by algorithm and described in text and in Table 1. The colors indicate the number of mainshocks for each sequence (the three sequences consisting of two mainshocks are shown in light green, and so on; see histogram in inset).

4. Simulation of the Seismicity

By means of a newly developed version of our physically based simulation code Console et al. [22] and references therein), we compiled synthetic earthquake catalogs lasting 100,000 years for events of magnitude ≥ 4.2 within the polygonal area depicted in Figure 1. In this version, no application is performed on the State and Rate formulation, but we adopted an enhanced role of the static Coulomb stress transfer between every ruptured element of the fault model and all the other elements in the surrounding faults. In this new version of the code, the magnitude distribution of the simulated catalog is controlled by two free parameters to be selected by the user (Console et al. [7,23]):

- The strength–reduction coefficient (S–R); this coefficient controls the growth of an initiated rupture, reducing the strength that must be exceeded for rupturing new elements of the expanding rupture, as a proxy of weakening mechanism;
- The aspect–ratio coefficient (A–R); this coefficient limits the progress of strength reduction if the ruptured area exceeds a given number of times the square of the width of the rupturing fault system, discouraging rupture propagation over very long distances.

The seismogenic model adopted in the simulation algorithm is depicted in Figure 1, and the slip rates assumed for each fault segment are the highest values of the range reported by the DISS database (DISS Working Group [9]; Table S1 of the Supplementary Material).

We carried out a set of tests to investigate the effect that the two above described free parameters have on the magnitude distribution of the output catalogs, letting the S–R parameter assume the values 0.1, 0.2, and 0.3, and the A–R parameter the values 2, 5, and 10, respectively. The results of these tests are reported in Figure S1 of the Supplementary Material. Each 100,000 years catalog was divided in 270 groups of 370 years (with the purpose of simulating many instances of the real catalog), counting the number of multiplets contained in each of them. Table 3 reports the averages and the standard deviation for the 270 elements population. Then, for each of the nine cases, we carried out the same analysis on 50 randomized catalogs obtained from the 100,000 years original ones by shuffling the origin time of all earthquakes by a random permutation. Finally, the average and the standard deviation of the ratios between the total number of multiplets in the original catalogs and those obtained from the respective randomized catalogs was computed (Table 4).

On the basis of the above-mentioned tests, although the largest number of multiplets is provided by the couple of parameters 0.1 and 2, we chose the simulation obtained with the values 0.2 and 10 for the S–R and A–R parameters, respectively, which gives the largest ratio of multiplets. Figure 3 shows the results of this simulation with the 13,845 earthquakes having $M_w \geq 5.0$, evidencing the fault segments where the number of simulated earthquakes is higher. Our simulation algorithm does not produce any seismic activity outside the borders of the faults considered in the seismogenic model.

A comparison of seismic features detected in the CPTI15 catalog, in the time interval 1650–2020, and the 100,000 years simulated catalog for the study area is shown in Table 5. For example, in this table we can compare the rate of earthquakes with $M_w \geq 5.0$ in the simulated catalog (0.138/yr) with the corresponding rate of earthquakes with $M_w \geq 5.0$ in the real catalog (0.573/yr). This circumstance is justified by the adoption of relatively high values of the S–R and A–R free parameters, which favor the growth of nucleated ruptures and accordingly produce a relatively large quantity of strong earthquakes. Moreover, we should take into account the fact that the source model adopted in our simulation does not include the numerous small sources, capable of producing only $M_w \leq 5.5$ earthquakes.

Table 5 shows a comparison of seismic features detected in the CPTI15 catalog, in the time interval 1650–2020, and the 100,000 years simulated catalog for the study area. In Figure 4, we show the Magnitude–Frequency Distribution (MFD) of the simulated catalog, compared with that of the 1650–2020 CPTI catalog for events above the completeness threshold magnitude of 5.0.

Another metric for comparing our simulations with the real observations is given by the numbers of multiplets counted in the same time interval of 370 years (third line of Table 5) and the mean ratio between these numbers and the corresponding numbers calculated on a set of randomisations (last line of Table 5): these randomisations should effectively destroy the presence of clustering relation among the events. The obtained mean number could in our opinion represent the degree of "clustering" of the catalogs. The value of the ratio systematically greater than one supports the hypothesis that the occurrence of sequences containing multiple mainshocks is not just a casual circumstance. Even if this procedure can give different results changing internal criteria, these criteria are not changed while applying the procedure to the two catalogs to be compared. Figure 5 shows the distribution of the ratios between the number of multiplets identified in the 100,000 years synthetic catalog and 500 randomizations of the same catalog. The average ratio is 2.13 ± 0.24, which denotes a good agreement between the production of multiplets of the simulated catalog with respect to that of the observations (see also Table 5).

In Figure 4, we show the MFD of the simulated catalog, compared with that of the 1650–2020 CPTI catalog for events above the completeness threshold magnitude of 5.0. This figure shows that the MFD of the simulated catalog does not follow a straight line as expected according to the Gutenberg–Richter law, but exhibits a change in its slope in the magnitude range $5.7 \leq M_w \leq 7.0$, where the b-value decreases dramatically. This circumstance is again due to the selection of the S–R and A–R free parameters, and the boostered role of the Coulomb stress transfer adopted in this particular study, which enhances the growth of nucleated ruptures, producing a sort of characteristic earthquake model.

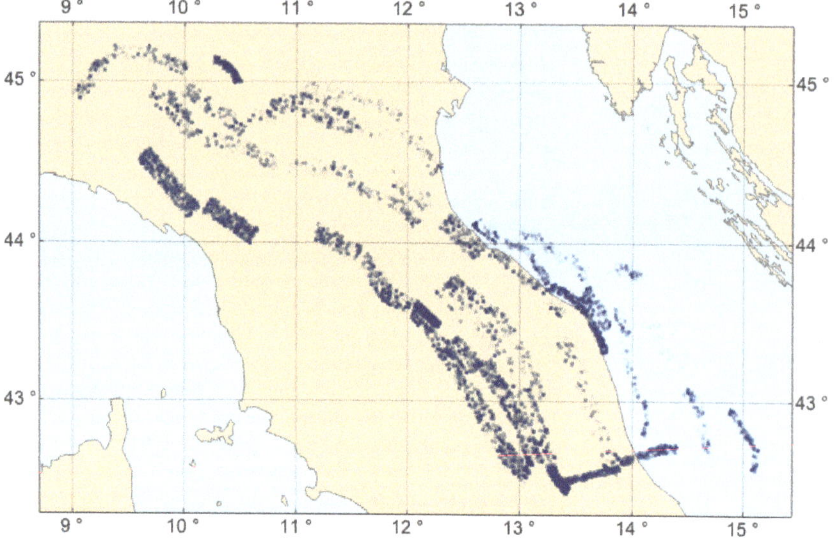

Figure 3. Map of 13,845 simulated earthquakes with $M_w \geq 5.0$, obtained from 100,000 years simulation. Point opacity is proportional to number of epicenters reported in output synthetic catalog in each cell of fault.

We also performed a comparison between the annual seismic moment rate released by the earthquakes of the simulated catalog and the observed ones. Adopting Hanks and Kanamori [24] magnitude–seismic moment conversion formula, we computed the total seismic moment released in the simulated catalog of 100,000 years. The sum is equal to $0.518 \cdot 10^{22}$ Nm, i.e., a seismic moment rate of $0.518 \cdot 10^{17}$ Nm/year. In a similar way, we computed the seismic moment of all earthquakes of M > 5.0 listed in the observational catalog from 1650 to 2020. A value of $1.35 \cdot 10^{19}$ Nm is acquired, implying a seismic moment rate of $0.365 \cdot 10^{17}$ Nm/year (Table 5). In conclusion, the seismic moment rate released by the simulated catalog is about 1.4 times larger than that of the observed seismicity. This could be explained by the uncertainties in the slip-rate values assumed in our seismogenic model.

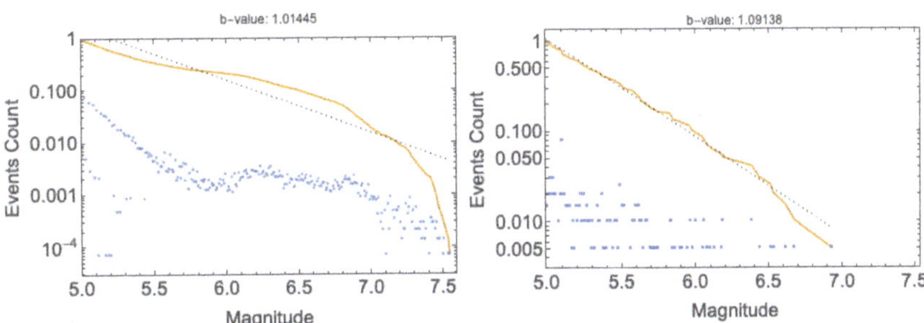

Figure 4. Cumulative (yellow line) and density (blue dots) Magnitude–Frequency Distributions (MFD) of $M_w \geq 5.0$ earthquakes of 100,000 years simulated (**left panel**) and observed (**right panel**; CPTI15 from 1650 to 2020 AD) catalogs. Straight dotted lines show best-fit of cumulative distributions.

We should also take into account the limited size of the earthquake catalog considered in the comparison of the observed seismic moment rate with that obtained from simulations. In fact, the duration of the 1650–2020 catalog (370 years) is shorter than the recurrence time on any of the fault segments reported in Table S1 and Figure 1. It is reasonable to hypothesize that this time window, upon which 22 events with $M_w \geq 6.0$ have occurred, was characterized by a moderate seismic activity in our study area, without a significant contribution of large magnitude events. In contrast with that situation, in the 17th and 18th centuries large magnitude earthquakes occurred in central-southern Italy, outside our study area.

In the same way as we prepared Figure 2, we also plot in Figure 6 the epicentres of the 100,000 years simulated catalog with $M_w \geq 5.0$. The comparison with Figure 2 shows that the simulated catalog is characterized by a scarce presence of sequences with a number of mainshocks larger than 2.

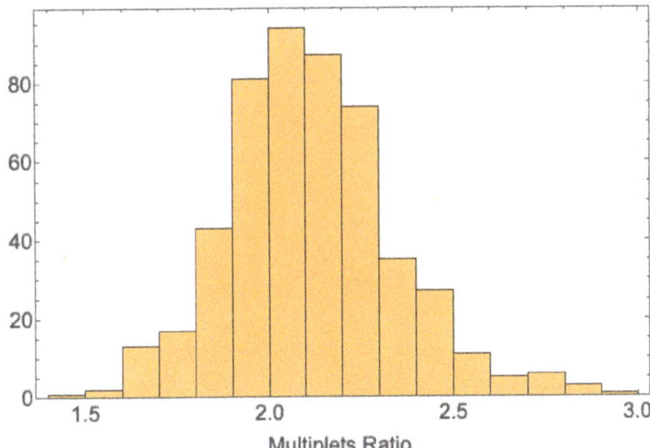

Figure 5. Histogram of ratios between number of multiplets identified in 100,000 years simulated catalog and 500 randomizations of same catalog.

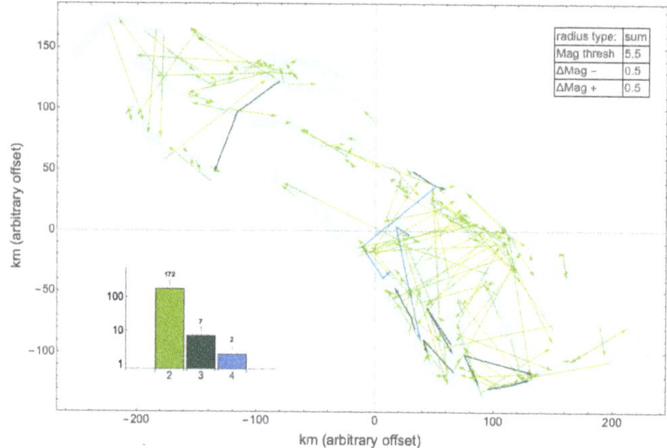

Figure 6. Representation of sequences with multiple mainshocks in 100,000 years simulated catalog, with $M_w \geq 5.0$. Colored lines show multiple mainshocks recognized by algorithm described in text, with colors indicating respective number of earthquakes for each of them (see histogram in log scale in the inset).

Table 3. Average number of multiplets in 100,000 years simulated catalogs in groups of 370 years.

Free Parameters	S–R = 0.1	S–R = 0.2	S–R = 0.3
A–R = 2	1.30 ± 0.07	1.10 ± 0.06	0.84 ± 0.05
A–R = 5	0.92 ± 0.05	0.71 ± 0.05	0.56 ± 0.04
A–R = 10	0.74 ± 0.05	0.65 ± 0.05	0.49 ± 0.04

Table 4. Ratio between total number of multiplets in original 100,000 years simulated catalogs and average of respective randomized catalogs.

Free Parameters	S–R = 0.1	S–R = 0.2	S–R = 0.3
A–R = 2	1.83 ± 0.11	1.77 ± 0.13	1.56 ± 0.11
A–R = 5	2.10 ± 0.20	1.94 ± 0.21	1.88 ± 0.18
A–R = 10	2.00 ± 0.18	2.13 ± 0.24	1.96 ± 0.23

Table 5. A comparison of seismic features detected in CPTI15 1650–2020 catalog and 100,000 years simulated catalog (S–R = 0.2 and A–R = 10) for study area.

Seismic Features	CPTI15	Simulation
Number of events of $M \geq 5.0$ per year	0.573	0.138
Seismic moment released per year (Nm)	$0.365 \cdot 10^{17}$	$0.518 \cdot 10^{17}$
Number of multiplets in 370 years	8	0.65 ± 0.05
Average number of multiplets in 370 years in the randomized catalogs	4.63 ± 1.88	0.32 ± 0.03
Average ratio of the number of multiplets between the original and randomized catalogs	2.17 ± 1.37	2.13 ± 0.24

5. Long- and Short-Term Features of the Simulated Seismicity

A detailed analysis of the simulated 100,000 years catalog allows the detection of interesting spatiotemporal features showing similarities with analog features existing in the observations.

The following stacking procedure was adopted to highlight if systematic and coherent time features occur before or after "strong" earthquakes: (1) We take into account earthquakes of the simulated catalog with a magnitude greater than 5.2; (2) for each of those events, occurring at time t_i, a time interval around it $(t_i - \Delta t, t_i + \Delta t)$ is considered and subgroups of events falling inside that interval and with an epicentral distance less than Δr are added to a stacking list. Their occurrence times are stored as counted relative to t_i, i.e, with times ranging from $-\Delta t$ to $+\Delta t$; (3) Once the stacking list was filled with all subgroups times, the resulting $(-\Delta t, \Delta t)$ interval is divided into a proper number of bins and events occurrences for any of the bins are counted and reported in the scatter plots.

Long-term seismicity patterns before and after a mainshock are shown in Figure 7. This figure shows the stacked number of $M_w \geq 4.2$ earthquakes that preceded and followed an $M_w \geq 5.2$ earthquake within an epicentral distance of 20 km. Here, we may note an acceleration of seismic activity some centuries before a mainshock, a modest quiescence starting 50 years before the mainshock and a strong aftershock occurrence in the following five years. After this aftershock phase, a trend of long-term quiescence recovering in some centuries is noted. In Figure S2 of the Supplementary Material we report the same kind of plots for all the nine combinations of free parameters, showing the same trends, with minor variations.

As far as the short-term features are concerned, a clear foreshock and aftershock pattern of the duration of some weeks before and after a magnitude $M_w \geq 5.2$ event is visible in the stacking plot of Figure 8a. With the same time scale, Figure 8b shows a clear trend of b-value decreasing before a mainshock of $M_w \geq 5.2$ and recovering to the average value just after it. Note that the large scattering of the b-values has no real physical meaning, but it is simply due to the limited number of earthquakes on which the b-value is calculated. However, this scattering is much smaller just before and after the mainshocks, when the earthquakes rate is much larger. In Figure S3 of the Supplementary Material, we report the same kind of plots for all the nine combinations of free parameters, showing the same trends, with minor variations. This feature was observed in natural sequences as, for instance, Montuori et al. [25], Papadopoulos et al. [26], Gulia and Wiemer [27].

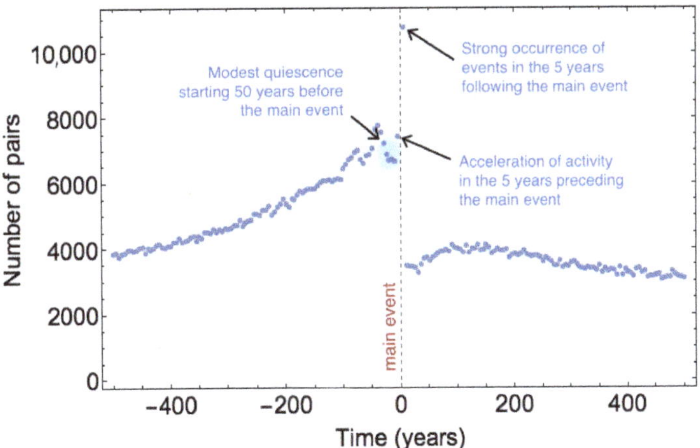

Figure 7. Stacked number of $M_w \geq 4.2$ earthquakes that preceded and followed an $M_w \geq 5.2$ earthquake within an epicentral distance of 20 km in 100,000 years simulated catalog.

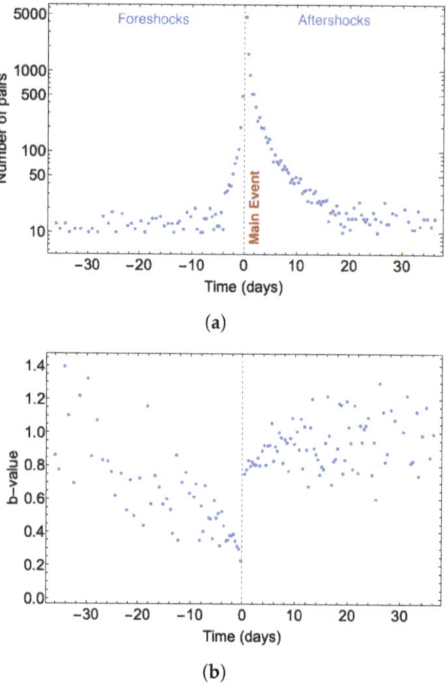

(a)

(b)

Figure 8. (**a**) Stacked number of $M_w \geq 4.2$ earthquakes that preceded and followed an $M_w \geq 5.2$ earthquake within an epicentral distance of 50 km in 100,000 years simulated catalog, zooming on a time scale spanning only 0.1 years (36.5 days); (**b**) average b-value in time bins of 0.365 days before and after an earthquake of $M_w \geq 5.2$ containing at least 10 events.

6. Conclusions

In this study, we assumed a definition of "multiplet" specifically tuned for the application to the seismicity of our study area. A computer code was developed for counting the number of multiplets detected applying such a definition to any earthquake catalog. For the CPTI15 1650–2020 catalog, the code detected eight multiplets, which is a number significantly higher than the average number of multiplets detected in the same way on a set of 500 data sets obtained randomizing the occurrence times of the original catalog (i.e., 4.63 ± 1.88 in Table 5). The result of this comparison supports the hypothesis that the occurrence of sequences containing multiple mainshocks is not just a casual circumstance. In this study, we also developed a new earthquake simulation code, paying particular attention to the enhancement of stress interaction among rupturing fault elements, and increased the number of multiplets in the simulated catalog. In this way, the number of multiplets detected by the above mentioned computer code on a 100,000-year simulated catalog (176) is 2.13 times higher than the average number of multiplets detected in the same way on a set of 500 randomized catalogs (Table 5). Besides the production of a significant number of multiplets, the simulated catalog exhibits long- and short-term spatiotemporal features that can be considered realistic imitations of those commonly observed in the real seismicity. We use a stacking procedure in which we compute the number of $M_w \geq 4.2$ events that preceded and followed an $M_w \geq 5.2$ earthquake in bins of five years considering the origin time at the time of every strong event in the 100 kyears simulated catalog. Our results related to northern and central Apennines show an acceleration of seismic activity some centuries before a mainshock, a modest quiescence starting 50 years before the mainshock and a strong aftershock occurrence in the following five years (Figure 7). In the same way, we analyzed the short term patterns in periods of about one month before and after every $M_w \geq 5.2$ earthquake. Our results confirm the capacity of the simulator

code to reproduce typical foreshocks-aftershocks sequences (Figure 8a). Additionally, in the simulated catalog for northern and central Apennines, the average b-values show a decrease lasting a few weeks before the strong earthquakes, followed by an instantaneous increase at the time of the earthquakes (Figure 8b). This pattern was observed by Montuori et al. [25], Papadopoulos et al. [26], Gulia and Wiemer [27] in real earthquake sequences.

Supplementary Materials: The following items are available online at https://www.mdpi.com/article/10.3390/app12042062/s1, Table S1: Geometric and kinematic parameters of the 198 quadrilateral fault segments derived from the 43 Composite Seismogenic Sources (CSS) of DISS v. 3.3.0. Figure S1: Cumulative and density magnitude–frequency distributions of $M_w \geq 4.2$ earthquake simulated catalogs, for the nine combinations of free parameters of Table 3 considered in this study. The straight dotted lines show the best-fit Gutenberg–Richter distributions. Figure S2: Stacked number of $M_w \geq 4.2$ earthquakes that preceded and followed an $M_w \geq 5.2$ earthquake within an epicentral distance of 20 km in the 100,000 years simulated catalog, for the nine combinations of free parameters of Table 3 considered in this study. Figure S3: (a) Stacked number of $M_w \geq 4.2$ earthquakes that preceded and followed up to 0.1 years (36.5 days) an $M_w \geq 5.2$ earthquake within an epicentral distance of 50 km in the 100,000 years simulated catalog. (b) b-value in the time bins of 0.365 days before and after an earthquake of $M_w \geq 5.2$ containing at least 10 events, for the nine combinations of free parameters of Table 3 considered in this study. Figure S4: Sketch of a quadrilateral fault segment.

Author Contributions: Conceptualization, R.C. (Rodolfo Console), P.V. and R.C. (Roberto Carluccio); methodology, R.C. (Rodolfo Console) and R.C. (Roberto Carluccio); software, R.C. (Roberto Carluccio) and R.C. (Rodolfo Console); validation, P.V., R.C. (Rodolfo Console) and R.C. (Roberto Carluccio); resources, P.V.; writing, R.C. (Rodolfo Console), R.C. (Roberto Carluccio) and P.V. All authors have read and agreed to the published version of the manuscript.

Funding: This research received no external funding

Institutional Review Board Statement: Not applicable.

Informed Consent Statement: Not applicable.

Data Availability Statement: The Italian Parametric Earthquake Catalogue (CPTI15) can be found at https://emidius.mi.ingv.it/CPTI15-DBMI15/ (accessed on 25 September 2021), the Database of Individual Seismogenic Sources (DISS) can be found at http://diss.ingv.it (accessed on 16 December 2021).

Conflicts of Interest: The authors declare no conflict of interest.

References

1. Rhoades, D.A.; Evison, F.F. Long-range earthquake forecasting with every earthquake a precursor according to scale. *Pure Appl. Geophys.* **2004**, *161*, 47–72. [CrossRef]
2. Rhoades, D.A.; Evison, F.F. Test of the EEPAS forecasting model on the Japan earthquake catalog. *Pure Appl. Geophys.* **2005**, *162*, 1271–1290. [CrossRef]
3. Rhoades, D.; Evison, F. The EEPAS forecasting model and the probability of moderate-to-large earthquakes in central Japan. *Tectonophysics* **2006**, *417*, 119–130. [CrossRef]
4. Evison, F.; Rhoades, D. The precursory earthquake swarm in New Zealand: Hypothesis tests. *N. Z. J. Geol. Geophys.* **1997**, *40*, 537–547. [CrossRef]
5. Gardner, J.; Knopoff, L. Is the sequence of earthquakes in Southern California, with aftershocks removed, Poissonian? *Bull. Seismol. Soc. Am.* **1974**, *64*, 1363–1367. [CrossRef]
6. Gentili, S.; Di Giovambattista, R. Forecasting strong aftershocks in earthquake clusters from northeastern Italy and western Slovenia. *Phys. Earth Planet. Inter.* **2020**, *303*, 106483. [CrossRef]
7. Console, R.; Murru, M.; Vannoli, P.; Carluccio, R.; Taroni, M.; Falcone, G. Physics-based simulation of sequences with multiple main shocks in Central Italy. *Geophys. J. Int.* **2020**, *223*, 526–542. [CrossRef]
8. Console, R.; Vannoli, P.; Carluccio, R. The seismicity of the Central Apennines (Italy) studied by means of a physics-based earthquake simulator. *Geophys. J. Int.* **2018**, *212*, 916–929. [CrossRef]
9. DISS Working Group. Database of Individual Seismogenic Sources (DISS), Version 3.3.0: A Compilation of Potential Sources for Earthquakes Larger Than M 5.5 in Italy and Surrounding Areas. Available online: https://doi.org/10.13127/diss3.3.0 (accessed on 16 December 2021).
10. Rovida, A.; Locati, M.; Camassi, R.; Lolli, B.; Gasperini, P.; Antonucci, A. *The Italian Earthquake Catalogue CPTI15—Version 3.0*; Istituto Nazionale di Geofisica e Vulcanologia (INGV): Rome, Italy, 2021.

11. Pondrelli, S.; Visini, F.; Rovida, A.; D'Amico, V.; Pace, B.; Meletti, C. Style of faulting of expected earthquakes in Italy as an input for seismic hazard modeling. *Nat. Hazards Earth Syst. Sci.* **2020**, *20*, 3577–3592. [CrossRef]

12. Mariucci, M.T.; Montone, P. Database of Italian present-day stress indicators, IPSI 1.4. *Sci. Data* **2020**, *7*, 1–11. [CrossRef]

13. Devoti, R.; d'Agostino, N.; Serpelloni, E.; Pietrantonio, G.; Riguzzi, F.; Avallone, A.; Cavaliere, A.; Cheloni, D.; Cecere, G.; d'Ambrosio, C.; et al. A combined velocity field of the Mediterranean region. *Ann. Geophys.* **2017**, *60*, S0215, doi:10.4401/ag-7059. [CrossRef]

14. Vannoli, P.; Burrato, P.; Fracassi, U.; Valensise, G. A fresh look at the seismotectonics of the Abruzzi (Central Apennines) following the 6 April 2009 L'Aquila earthquake (Mw 6.3). *Ital. J. Geosci.* **2012**, *131*, 309–329. [CrossRef]

15. Michele, M.; Chiaraluce, L.; Di Stefano, R.; Waldhauser, F. Fine-scale structure of the 2016–2017 Central Italy seismic sequence from data recorded at the Italian National Network. *J. Geophys. Res. Solid Earth* **2020**, *125*, e2019JB018440. [CrossRef]

16. Basili, R.; Valensise, G.; Vannoli, P.; Burrato, P.; Fracassi, U.; Mariano, S.; Tiberti, M.M.; Boschi, E. The Database of Individual Seismogenic Sources (DISS), version 3: summarizing 20 years of research on Italy's earthquake geology. *Tectonophysics* **2008**, *453*, 20–43. [CrossRef]

17. Bonini, L.; Basili, R.; Burrato, P.; Cannelli, V.; Fracassi, U.; Maesano, F.E.; Melini, D.; Tarabusi, G.; Tiberti, M.M.; Vannoli, P.; et al. Testing different tectonic models for the source of the Mw 6.5, 30 October 2016, Norcia earthquake (central Italy): A youthful normal fault, or negative inversion of an old thrust? *Tectonics* **2019**, *38*, 990–1017. [CrossRef]

18. Di Bucci, D.; Buttinelli, M.; D'Ambrogi, C.; Scrocca, D.; Anzidei, M.; Basili, R.; Bigi, S.; Bignami, C.; Bonini, L.; Bonomo, R.; et al. RETRACE-3D project: A multidisciplinary collaboration to build a crustal model for the 2016–2018 central Italy seismic sequence. *Boll. Geofis. Teor. Appl.* **2021**, *62*, 1–18. [CrossRef]

19. Vannoli, P.; Vannucci, G.; Bernardi, F.; Palombo, B.; Ferrari, G. The source of the 30 October 1930 Mw 5.8 Senigallia (Central Italy) earthquake: A convergent solution from instrumental, macroseismic, and geological data. *Bull. Seismol. Soc. Am.* **2015**, *105*, 1548–1561. [CrossRef]

20. Vannoli, P.; Burrato, P.; Valensise, G. The seismotectonics of the Po Plain (northern Italy): Tectonic diversity in a blind faulting domain. *Pure Appl. Geophys.* **2015**, *172*, 1105–1142. [CrossRef]

21. Zampieri, D.; Vannoli, P.; Burrato, P. Geodynamic and seismotectonic model of a long-lived transverse structure: The Schio-Vicenza Fault System (NE Italy). *Solid Earth Discuss.* **2021**, *12*, 1967–1986. [CrossRef]

22. Console, R.; Carluccio, R.; Murru, M.; Papadimitriou, E.; Karakostas, V. Physics-Based Simulation of Spatiotemporal Patterns of Earthquakes in the Corinth Gulf, Greece, Fault System. *Bull. Seismol. Soc. Am.* **2022**, *112-1*, 98–117. [CrossRef]

23. Console, R.; Nardi, A.; Carluccio, R.; Murru, M.; Falcone, G.; Parsons, T. A physics-based earthquake simulator; its application to seismic hazard assessment in Calabria (Southern Italy) region. *Acta Geophys.* **2017**, *65*, 243–257. [CrossRef]

24. Hanks, T.C.; Kanamori, H. A moment magnitude scale. *J. Geophys. Res. Solid Earth* **1979**, *84*, 2348–2350. [CrossRef]

25. Montuori, C.; Murru, M.; Falcone, G. Spatial variation of the b-value observed for the periods preceding and following the 24 August 2016, Amatrice earthquake (ML 6.0)(central Italy). *Annals Geophys.* **2016**, *59*. [CrossRef]

26. Papadopoulos, GA; Minadakis, G.; Orfanogiannaki, K. Short-term foreshocks; earthquake prediction. In *AGU Geophysical Monograph Series Book*, 1st ed.; American Geophysical Union: Washington, DC, USA, 2018; pp. 127–147.

27. Gulia, L.; Wiemer, S. Real-time discrimination of earthquake foreshocks; aftershocks. *Nature* **2019**, *574*, 193–199. [CrossRef]

MDPI
St. Alban-Anlage 66
4052 Basel
Switzerland
Tel. +41 61 683 77 34
Fax +41 61 302 89 18
www.mdpi.com

Applied Sciences Editorial Office
E-mail: applsci@mdpi.com
www.mdpi.com/journal/applsci